The Seventh Year
Industrial Civilization
in Transition

The Seventh Year
Industrial Civilization in Transition

by W. JACKSON DAVIS

W · W · NORTON & COMPANY · NEW YORK · LONDON

ACKNOWLEDGMENTS:

Figures 3.1, 3.2, 5.7, 6.3, 6.5, 7.5, 7.6, 7.7, 7.8, 9.4, and Table 3.1; © 1974, © 1974, © 1976, © 1976, © 1977, © 1976, ⓐ 1976, © 1976, © 1976, © 1976, and © 1974 respectively by the American Association for the Advancement of Science.

Figures 4.2, 13.3, and 13.4; © 1977, © 1976, and © 1978 respectively by The New York Times Company. Reprinted by permission.

Figure 6.2 appears courtesy of ECOLOGIST.

Figures 6.4, 9.7, 10.6, and 11.1; © 1978, © 1976, © 1976, and © 1974 by Scientific American, Inc. All rights reserved.

Figures 7.3 and 7.4 appear courtesy of University Publications, Blacksburg, VA.

Figures 9.3, 9.6, 9.9, 9.10, and 10.2, © 1974 by J. S. Steinhart and C. E. Steinhart.

Figure 10.5 appears courtesy of the International Institute for Environment and Development.

Figure 11.4 appears courtesy of the Society for Experimental Biology.

Library of Congress Cataloging in Publication Data
Davis, William Jackson, 1942–
The seventh year.
Includes bibliographical references and index.
1. Technology and civilization.   2. Civilization,
Modern—1950–   3. Natural resources.   I. Title.
CB478.D275   1979   909.82   79-1324
ISBN 0-393-05693-7
ISBN 0-393-09027-2 pbk.

2 3 4 5 6 7 8 9 0

For Kimberly Jane

CONTENTS

PREFACE AND ACKNOWLEDGMENTS   xiii

INTRODUCTION   xv
   Predicting the Future   xvi
   On "Optimism"   xviii
   On Preparation   xviii
   Notes   xix

PART I: FOUNDATIONS   1

Chapter 1: ENERGY   3
   What is Energy?   4
   Kinetic and Potential Energy   4
   How Is Energy Stored?   5
   The Law of Energy Conservation   6
   The Law of Energy Degradation   7
   Heat Does Work   8
   Thermal Efficiency   10
   Where Does Earth Get Its Energy?   10
   What Happens to Solar Energy on Earth?   11
   Entropy and Life   14
   Notes   15
   Additional Resources   15

Chapter 2: THE EVOLUTION OF INDUSTRIALISM   17
   How Is Energy Made Available to Living Forms? 17
   What Happens to Green Plants?   18
   What Happens to Herbivores?   19
   What Happens to Carnivores?   19
   Ecology Equals Thermodynamics   20

Cultural Evolution and Natural Law   20
The Industrial Age   24
The Lessons of Ecocultural History   26
The Role of Positive Feedback   29
Notes   31

Chapter 3: THE ENERGY BASIS OF INDUSTRIALISM   33
Sources of Energy for U.S. Industrialism   34
Uses of Energy by Economic Sector   35
Fossil-Fuel Production Cycles: The Exponential Model   37
The Last Easter Egg   40
Fossil-Fuel Production Cycles: The Bell-Shaped Model   42
Hubbert Versus the Giants   47
King Hubbert's Hump   49
Short-Term Solutions   50
Notes   52

Chapter 4: ALTERNATIVES TO CONVENTIONAL FOSSIL FUELS   54
Synthetic Fuels   54
Oil Shale and Tar Sands   56
Nuclear Power   58
Energy Income   67
Energy Conservation   70
The Future of Energy   72
Notes   73

Chapter 5: NATURAL RESOURCES, ECOLOGY, AND ECONOMY   77
What Is a Natural Resource?   77
The Classification of Natural Resources   78
The Ecology of Natural Resources   81
Economics and the Resource Cycle   86
Keynesian Economics   94
Toward a New Economic Paradigm   94
Notes   96

Chapter 6: PRIMARY NATURAL RESOURCES   98
The Climate   98
Fresh Water   103
The Oceans   106
The Atmosphere   109
Land   112
The Future of Primary Natural Resources   115
Notes   116

Chapter 7: SECONDARY NATURAL RESOURCES 121
Animal Resources 121
The Forest Resource 123
Mineral Resources 126
The Depletion of Metals 127
Cornucopia Revisited 132
The Age of Substitution 137
Recycling 137
Foreign Dependence 138
Metals and the Future 140
Notes 141

Chapter 8: THE IMMEDIATE LIMITS TO GROWTH 144
The Nature of Limited Processes 144
Environmental Limits and Negative Feedback 146
The "Lead Time" Limit 148
Sociological Limits 150
Political Limits 151
Institutional Homeostasis 152
Economic Limits 153
The Limits of Technology 156
Limits and the Principle of Interrelatedness 159
Notes 162

PART II: CONSEQUENCES 165

Chapter 9: THE STAFF OF LIFE 167
Roots: The Agricultural Revolution 168
The Capital Phase of U.S. Agriculture 169
The Energy Phase of U.S. Agriculture 170
Farm Methodology 172
Farm Population 174
Farm Size 174
Corporate Agribusiness 175
The Evolution of the Supercrop 176
The Age of the Monocrop 178
The Ecology of Monocropping 179
Pesticides: A Case Study in Diminishing Returns 180
The Green Revolution 181
Energy Equals Food 181
Diminishing Returns 185
Notes 187

Chapter 10: THE SEEDS OF CHANGE   191
  The Agricultural Equation   192
  The Decentralization of Agriculture   193
  The Decline of Food Processing   194
  Changing What We Eat   194
  Enriching the Earth   197
  The Despecialization of Agriculture   198
  Demechanization and the Rise of Animal Power   199
  Economics: The Engine of Change   200
  A Return to the Past?   202
  Models from the Present   203
  Bridges to the Future   205
  Looking Ahead   210
  Notes   211

Chapter 11: POPULATING THE EARTH   215
  The History of Human Population   216
  Population Growth   218
  The Theory of Demographic Transition   218
  The Age Structure of Populations   220
  The Future of Global Population   223
  The Concept of Carrying Capacity   224
  Global Carrying Capacity   225
  Global Population Densities   226
  The Urbanization of the Human Population   229
  Urban Population Densities   231
  The Metabolism of the City   232
  The Consequences of Carrying Capacity   233
  Notes   234

Chapter 12: THE DECLINE OF THE CITY   237
  The Plight of the City   237
  The Human Cost of Urban Decline: A Real-Life Scenario   248
  Forecasting the Future of the City   251
  Energy, Resources, and National Urban Policy   252
  Notes   255

Chapter 13: VISIONS   262
  Living Patterns   264
  The Revaluation of Work   268
  The Restructuring of Schooling   271

The Decentralization of Political Institutions 273
Culture, Values, and Ethics 276
The Birth of Eco-Logic 279
The Dawn Approaches 280
Notes 282
INDEX 287

PREFACE AND ACKNOWLEDGMENTS

This book is based on an undergraduate course that I teach at the University of California at Santa Cruz on the past, present, and possible futures of industrial civilization. The course is part of the collegiate teaching program at the University, to which I am grateful for its policy of encouraging interdisciplinary excursions such as this. I am especially indebted to my friends at W. W. Norton, including Joseph B. Janson II, Ethelbert Nevin II, Nathan L. Wilbur, James D. Jordan, Mary Shuford, Diane Nish and others, for helpful advice and for enabling prompt publication of these lectures in the form of this book. I am grateful to the several reviewers, including Edward Cornish, Paul Ehrlich, A. M. Freeman III, Suzanne Keller, and especially Edward J. Kormondy for competent and perceptive criticisms. I thank Clark Carroll for preparing the illustrations, and Polly Brodecky and Dale Lewis for typing the manuscript. The love and generosity of Ruth P. Carter and the late Richard Carter greatly facilitated the writing of this book.

I owe the greatest debt to my colleague and wife, Kimberly Jane Davis. She provided not only warm encouragement and support, but also intellectual stimulation and expert editing. Several of the ideas herein were forged by the fireside in joint conversation that filled many a long winter evening.

WJD

Santa Cruz, California
December 1978

INTRODUCTION

In ancient Egypt, before the birth of Christ, the young Pharaoh Amenhotep dreamt a prophetic dream that was to change the course of history. In his dream he saw seven sleek and healthy cows emerge from the waters of the Nile River to feed upon its lush banks. Within moments they were each devoured by seven starved and haggard cows that followed. Haunted by the dream, the pharaoh called upon his wisest priests, but none could fathom its meaning until Joseph was summoned from his dungeon. Joseph accurately interpreted the dream as a prophecy of God, foretelling of seven years of plenty to be followed by seven years of famine throughout the lands of Egypt. The pharaoh appointed Joseph to preside over the preparations.[1]

Now, more than 20 centuries later, the pharaoh's dream is a distant memory, and the surface of the earth has been transformed. The Industrial Revolution has expanded the wealth of nations beyond the fondest hopes of Adam Smith; human population has multiplied beyond the morbid expectations of Thomas Malthus; and each generation of human beings has produced and accumulated knowledge beyond the combined imaginations of all preceding generations. It has been a time of plenty.

But in the midst of plenty there grows a pervasive unease. Beneath the facade of gleaming jet airliners, soaring skyscrapers, and climate-controlled shopping centers, we sense increasingly that something is desperately out of control. We are specks on the river of history, swept toward a sharp bend that will carry us into the unfamiliar. For more than two centuries industrial civilization has borrowed on credit from nature, and now an enormous debt is due. Industrialism has entered its seventh year of plenty.

This book is in part a chronicle of the future. It is a study of a period

in history that I believe will arrive within the lifetime of people now alive.
It is a study of the ending of the Industrial Age and the birth of a new
human civilization whose outlines can be dimly perceived. Birth is seldom without pain, but my intent is not to "bruise the heart . . . with the
traditional announcements of horrible misery to come."[2] Rather, my intent is to prepare the heart with a vision of a future that could be.

Predicting the Future

Futurists—people who study the future—are generally repelled by the
term "predict," for in times past prediction was the province of the seer,
the sorcerer, and the religious prophet. But when we hurl a stone sky-
ward we are not hesitant to predict that it will return to earth. Likewise,
we can predict with assurance that the sun will rise tomorrow and that
Halley's comet will pass closest to our sun again on February 9, 1986, at 4
P.M. Eastern Standard Time.[3] Knowledge generated largely during the
Industrial Age has begun to place the art of prediction on the footing of a
science, with the result that the field of futuristics has blossomed in the
last two decades.[4]

The Role of Pattern

But lest our successes blind us to our limits, let us recall that we cannot even yet predict the weather reliably, let alone our own behavior.
Why is one event so readily predictable and another so elusive? A partial
answer lies in pattern. We can predict the coming of a comet in part
because the natural sciences have deciphered the pattern of celestial
mechanics. We will make use of this lesson as we analyze the future of industrial civilization. In the first half of this book we will examine industrialism for its past and present patterns, relying heavily on the perspective supplied by the natural sciences, and especially ecology.

The Role of Probability

But there is more to prediction than pattern. We recognize that the
future can never be predicted with certainty, but only with a given probability. That branch of mathematics known as probability theory counsels
that the reliability of prediction is greatest when the number of variables
is least. We can predict the movements of the heavenly bodies with such

precision in part because their wanderings are determined by a small number of variables, such as mass and gravity. Likewise, predicting the future of industrialism requires that we first limit and identify the critical variables. Accordingly, in the first half of this book we will see how three variables—energy, natural resources, and technology—are interrelated and have been central to the affluent way of life that typifies modern industrial civilization.

The Role of Paradigm

But if prediction were shaped only by pattern and probability, then futurists with access to the same information and methodology ought to arrive at identical conclusions. Clearly they do not. Daniel Bell projects the evolution of industrialism to a technologic civilization founded on "theoretical knowledge" and run by an elite technical class, the "meritocracy."[5] In contrast, Robert Heilbroner foresees the collapse of business civilization and the growth of authoritarian political regimes.[6]

To what may we ascribe such contradictory conclusions? The answer, in a word, is *paradigm*. A paradigm, as defined by Kuhn, is a set of broadly shared assumptions that governs scientific inquiry.[7] Kuhn has recognized that the natural sciences are guided less by logic and objectivity than by the reigning paradigm. It is the paradigm that dictates which observation is important, which hypothesis merits attention, and which theory holds sway. Likewise, paradigms mold predictions about the future. In forecasting the future of industrialism, Bell operates by one paradigm, Heilbroner another. Bell explicitly assumes that energy and natural resources are effectively unlimited and that human beings are eminently rational. Heilbroner assumes that energy and resources are limited, and that democracy may be incapable of coping with the effects of their depletion.

Thus we shall draw a third lesson from the natural sciences and examine carefully the paradigm and assumptions on which this work is based. As detailed in the first half of the book, the general paradigm is developed largely from the science of ecology, and the central assumption is that energy and natural resources will decline in availability in the foreseeable future. In the second half of the book we will explore some implications of energy and resource scarcity for industrial civilization. I will develop the thesis that energy and resource "exhaustion" is unnecessary to transform our civilization radically: the high cost that attends depletion is a sufficient driving force. I will propose that this economic force may shortly propel industrial societies through a transition as fundamental as the one that ushered in the Industrial Revolution. Historians may come to see the Industrial Age as a brief but pivotal episode in cultural evolution, characterized primarily by the use of energy and resources to generate material knowledge. The coming transition can be seen as an engine of evolution, requiring humankind to employ this knowledge to establish new values and directions, and to forge a fresh and fundamentally different relationship with itself, nature, and the universe.

On "Optimism"

Mine is an optimistic thesis, but not in the conventional usage of the word. Our culture is so wedded to the material fruits of industrialism and to Western notions of progress that optimism has become synonymous with growth and affluence. But if the thesis of this work is correct, optimism for the future cannot be based realistically on the hope for continued material abundance. I believe the time of plenty is nearly past. I propose to return to Leibniz's original definition of optimism as the philosophy that "reality" is basically benign, regardless of the particular form it takes. It is then much easier to see the good in change, and to transcend the futile longing for the status quo.

I propose further that we view the coming transition in broad ecological perspective, for optimism is then unavoidable. Industrialism has enriched a small fraction of humanity and generated vast knowledge of the material universe. But as I will document, industrialism has also impoverished a segment of humanity, and deeply scarred this patient planet that has been given into our custody. The past benefits of industrialism are undeniable—but so also are the costs. I believe that the coming transition will give our species new values and fresh direction, and at the same time provide this magnificent earth a welcome respite, a precious time to cleanse its waters and heal its wounds. In short, I see the coming changes as beneficial to this earth and all the life forms it nourishes, including our own species. It is in this sense that I am optimistic.

On Preparation

We may imagine many paths to the future, some more painful than others. But if the assumptions underlying this work are correct—and time will provide an immediate test—then all paths lead to a common destination. My purpose here is not to catalog the catastrophes that may befall us along the way; these have been explored sufficiently by other writers. Rather, I seek to envision the destination, that we may better prepare for the journey. With other writers before me I believe it is time for the formulation of new individual and national motives, priorities, paradigms, and values.[8] The time is ripe for educators to instruct our children in new facts and futures. The time is ripe for economists to develop a post-Keynesian economic theory designed around the limits of the ecosystem. It is time for governments and governed alike to address seriously the prospect that energy and natural resources are approaching a period of ir-retrievable decline. It is time to incorporate this contingency into individual, local, and national policies. In short, it is time to prepare. I hope to persuade the reader that although the challenge is immense, the prospect is exhilarating and the opportunities abound.

When Joseph completed his interpretation of the pharoah's dream, His Majesty rejoiced that

> where there is no time, then of course one cannot take any; but we can,
> for before us lies a fullness of it. Seven years! That is the great thing: the
> fact to make us dance and rub our hands together.[9]

Likewise we may be thankful, as individuals and as a society, that we are granted time to choose a path. The path may be easier if we prepare; my purpose in writing this book is to contribute, however humbly, to the preparation.

NOTES

1. The story is told in Genesis 41, and is brought to vivid life in T. Mann, *Joseph and His Brothers* (New York: Knopf, 1958).

2. Mann, *Joseph and His Brothers*, p. 938.

3. D. K. Yeomans, "Comet Halley—The Orbital Motion," *Astronomical Journal* 82 (1977): 435–40.

4. E. Cornish, *The Study of the Future* (Washington, D.C.: World Future Society, 1977); World Future Society, *The Future: A Guide to Information Sources* (Washington, D.C.: World Future Society, 1977).

5. D. Bell, *The Coming of Post-Industrial Society* (New York: Basic Books, 1973).

6. R. Heilbroner, *The Human Prospect* (New York: W. W. Norton, 1974); id., *Business Civilization in Decline* (New York: W. W. Norton, 1976).

7. T. S. Kuhn, *The Structure of Scientific Revolutions* (Chicago: University of Chicago Press, 1962).

8. E. F. Schumaker, *Small Is Beautiful* (London: Blond and Briggs, 1973); W. Ophuls, *Ecology and the Politics of Scarcity* (San Francisco: W. H. Freeman, 1977).

9. Mann, *Joseph and His Brothers*, p. 949.

PART I
FOUNDATIONS

I believe that industrial civilization stands at the threshold of a period of accelerated cultural evolution that will culminate in a new and profoundly different civilization from the one we now know. I believe that this transformation will be propelled by the irreversible depletion of energy and natural resources; that it may occur within the lifetime of our children; and that the resulting civilization will be more advanced and satisfying than we might imagine.

To develop this thesis in a plausible manner requires background. For example, to appreciate the irreversibility of energy and resource depletion demands an acquaintance, however brief, with the laws of thermodynamics and other energy principles. Likewise, to accept that energy and resource depletion are feasible within our lifetimes requires a careful examination of the most recent data on use rates and remaining reserves of fossil fuels, as well as an investigation of possible alternatives. To comprehend fully the effects of energy and resource depletion requires understanding the links among energy, resources, and the economy. And to speculate meaningfully on the future of industrial civilization requires a broad understanding of its past and present.

These are among the topics of Part I. My intent is to furnish the foundation that is essential to developing the consequences of energy and resource depletion for the future—the subject of Part II.

Chapter 1
ENERGY

Everything is energy. The stars are energy; our thoughts are a form of energy; everything in our lives that we can touch or see is a form of energy; and industrialized societies such as our own depend for their existence on energy. By an industrialized society, I mean one whose economy and social organization are based directly on industrialism. Stripped to bare essentials, industrialism is the process by which human labor is amplified by large-scale energy conversion and directed toward the manufacture of goods from natural resources. We will frequently return to this definition, which can be summarized in the form of a chemical equation, with reactants on the left, products on the right, and energy as the driving force (Figure 1.1).

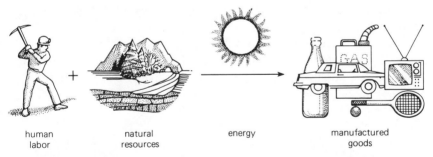

| human labor | natural resources | energy | manufactured goods |

1.1 The "industrial equation." This drawing schematically illustrates the process of industrialism, by which human labor is amplified by energy to convert natural resources to manufactured goods.

Industrialism of course produces more than just manufactured goods. It also produces science (knowledge of the material universe) and technology (application of this knowledge). Directly or indirectly, industrialism produces people, pollution, food, and values. We shall have oc-

3

casion to discuss all of these as we develop the thesis of this book, that industrialism is a brief and inherently transient step in the evolution of human culture.

But look again at the industrial equation (Figure 1.1) and note that energy occupies a central position. Energy is the prime mover of industrialism; it keeps the wheels of the system turning. To understand our industrial society and its future, and to see why our grandchildren may look upon the Industrial Age as history, let us begin at the beginning.

What Is Energy?

Stated in its simplest form, energy is the capacity for doing work. Physicists, with their inclination for precision, are fond of definitions. They define work not as an activity we exchange for a weekly paycheck, although this kind of work certainly includes some of their kind of work. Physicists define work as a product of two variables: a force applied to an object, and the distance that the object moves as a result of the force. Physicists thus measure work not only by an effort expended on its behalf, but also by the effect of that effort as reflected in movement produced. If we push on a boulder for several hours and it doesn't budge, we have not done the physicist's kind of work.

This definition of work is a flexible one, and its elements can be interchanged in illuminating ways. For example, if we see something move, we can be assured that work has been done and therefore energy expended. This form of the definition applies fully to industrialism; the movement of the turbines of an electric generator requires the expenditure of a good deal of energy. Similarly, the converse of our original definition is valid. That is, when energy is withdrawn, movement necessarily slows and eventually ceases, a statement that likewise applies fully to industrialism.

Kinetic and Potential Energy

Our definition of work can be seen in yet another light. That is, if a body is moving by definition it contains energy. Physicists call the energy contained in a moving body kinetic energy. The moving turbines of an electric generator have kinetic energy, which is converted by the machine to electricity. The fuel that set the turbines moving in the first place also contains energy, but this energy is stored in a nonmoving form as the potential for doing work. This kind of energy is termed, reasonably enough, potential energy.

How Is Energy Stored?

Energy takes many forms, most of which are familiar from everyday experience. Gravity, light, and electricity are examples of forms that energy may take. Three forms of energy, however, are especially relevant to industrial societies and their future: chemical energy, nuclear energy, and heat.

Chemical energy is stored in the fossil fuels on which industrial societies depend. These fuels are thought to have originated from plants, ancient swamps and forests buried by the ceaseless sliding movements of the earth and converted by different chemical reactions to coal, oil, and natural gas. The chemical energy in fossil fuels was thus put there in the first place by the process by which plants grow, namely, photosynthesis. By this process, green plants use sunlight to convert carbon dioxide in the atmosphere to simple natural sugar and oxygen (Figure 1.2).

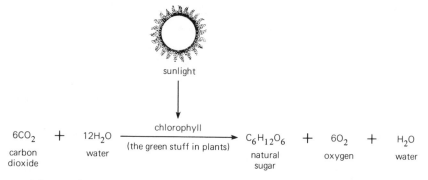

sunlight

$$6CO_2 \ + \ 12H_2O \xrightarrow[\text{(the green stuff in plants)}]{\text{chlorophyll}} C_6H_{12}O_6 \ + \ 6O_2 \ + \ H_2O$$

carbon dioxide   water   natural sugar   oxygen   water

1.2 Photosynthesis—the natural chemical process by which all fossil fuels were produced. Carbon dioxide and water are converted by green plants to natural sugar, oxygen, and water. The natural sugar is in turn chemically altered by nature to yield fossil fuels.

The simple sugars produced originally by photosynthesis still exist in fossil fuels, although in chemically modified and concentrated form. Most of the energy in fossil fuels is contained in the chemical bond that holds hydrogen and carbon together (the hydrocarbon bond). This energy can be extracted by breaking the bond, which is accomplished by burning, or oxidizing, the fossil fuels. The chemical reaction is basically the reverse of photosynthesis: the solar energy that was converted by photosynthesis to chemical energy comes out by oxidation as heat, which can be converted by suitable machines to work. When we ride the bus or drive to school or work, we are propelled in essence by sunshine that bathed the earth millions of years ago.

A second form of energy, increasingly important to industrial societies, is nuclear energy. It is the most concentrated known form of energy: more nuclear energy is stored in a marble than there is chemical energy in a truckload of coal. Nuclear energy is contained not in bonds between different atoms, as is chemical energy, but rather in the nuclei or central portion of the atoms themselves. Nuclear energy is the "glue" that holds matter together. Industrial societies can extract nuclear energy in two ways. The first, termed fission, involves splitting apart one atom into two smaller atoms which together weigh less than the original one. The difference in mass is realized as energy and is obtained as heat. The second way energy can be extracted from atomic nuclei is fusion. By this process, two atoms are forced together until they fuse to form a larger atom. The two summed are greater than the one, and again the difference is energy obtained as heat.

Heat, the crucial third form that energy takes, is a form of kinetic energy that results from the constant random movements of atoms. The faster these movements, the greater the kinetic energy and the hotter a thing feels. Both the chemical energy of fossil fuels and the nuclear energy in atoms must be converted to heat before they can do the work of industrialism. The rules that govern this conversion go by the somewhat imposing title of the laws of thermodynamics. These laws tell us why heating our homes with electricity is less efficient than heating with firewood, why eating meat is energetically less efficient than eating vegetables, and ultimately they even tell us why our industrial economies cannot grow indefinitely.

The Law of Energy Conservation

The first law of thermodynamics was discovered with extraordinarily simple apparatus and the native ingenuity of a 19th-century English physicist named James Prescott Joule (1818–1889). Joule was a brewer by calling, indulging himself in science when that now-honored profession was little more than the engaging hobby of the well-to-do. At the age of 22, he had the excellent sense to stir a tub of water with wooden paddles and periodically measure its temperature with a thermometer. He found that the work he performed in turning the paddles was expressed in the increased temperature of the water, from which he deduced that energy is neither created nor destroyed, but rather converted from one form to another. In Joule's case, chemical energy stored in the form of simple natural sugars (glucose) in the body was converted to mechanical energy by his muscles, and ultimately to heat.

The Law of Energy Degradation

If energy is neither created nor destroyed, the enterprising reader might well inquire exactly why we have an energy crisis. Why not recycle energy, making it do work again and again? This enticing possibility has excited the imagination of many great thinkers, who accordingly sought to discover the scientist's version the fountain of youth, the perpetual motion machine. Both fancies are foiled by the same villain—entropy. In Joule's historic experiment his energy was not lost, but merely transferred to the water. In the process, however, the chemical energy in glucose was dispersed in heat; that is, the organization or concentration of the energy declined. Likewise, when a bottle is broken, a ship is wrecked, or a perfectly ordered deck of cards is shuffled, matter is not lost; its organization merely becomes more random. Entropy is a measure of randomness: when organization decreases or energy is dispersed, entropy increases.

The second law of thermodynamics states that in any isolated (closed) system, entropy on the average increases. That is, the system is bound to become disordered. The universe itself may be such a closed system that is slowly running down. Crawley, who provides an excellent introductory account of energy principles, refers to entropy as "time's arrow,"[1] because entropy is what imparts to time its directionality, its sense of moving forward.

To illustrate the law of entropy, consider an example that is directly relevant to industrialism. A lump of coal is not, by conventional standards, an exceptional object. But if we could shrink to the size of an electron and enter the lump of coal, as Alice entered the looking-glass, we would marvel at a delicately crystalline universe, at galaxies arrayed in delightfully ordered lattices and at solar systems bound together in permanent if stochastic precision. It is small and humble, this lump of coal, but beneath its surface lies an extraordinarily organized miniature universe. This miniature universe was assembled over eons of geologic time by life processes that required still more eons to evolve. When the chemical energy invested in the coal by photosynthesis is released by oxidation, however, this miniature universe is disassembled in an instant and reduced to heat energy. The energy is not lost; it still exists somewhere in the form of the accelerated motion of atoms and molecules. The energy is, however, dispersed in space and time. In its inevitable way, entropy (randomness) has increased.

The second law teaches a lesson that will be recalled frequently in pages to come. Namely, the more ordered or concentrated a form of energy is, the higher is its "quality." Nuclear energy is as concentrated, we think, as energy can get. Fossil fuels are also "high-quality" energy sources because the energy put into them originally by photosynthesis has been concentrated by time. Solar flux on today's earth, while abundant, is a "low-quality" form of energy because it is so diffuse. As we shall detail in Chapter 4, it is difficult for people to concentrate sunlight enough to do much useful work, although, given time, plants do an effective job by means of photosynthesis.

Heat Does Work

What we really mean when we state that energy is concentrated or "high quality" is that it contains more prospective heat per unit volume, for ultimately it is heat that does the work of industrialism. To see why this is so, let us first rely on a simple abstraction and then extend the revealed concepts to the real world.

Imagine an object that is uniformly hot (or cold—it matters not), and imagine further a barrier right down its middle (Figure 1.3a). As long as the temperature on one side is the same as on the other, the molecules on one side rattle like a billion ping-pong balls against the barrier as frequently as those on the other side. It is a tug of war in reverse, with equal contestants; the equal but opposite forces exactly cancel, and the barrier is unmoved. With no movement there is by definition no work.

Now, let us enter the contest and heat one side of our imaginary object (Figure 1.3b). The molecules on that side move faster, so that in any given instant of time the barrier is bombarded by more molecules from that side. Net force is generated toward the cooler side and the barrier moves in that direction. With movement there is work.

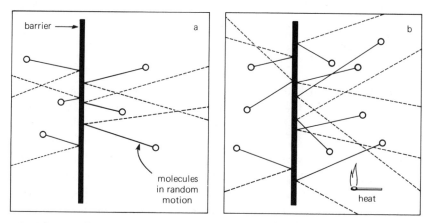

1.3 Schematic illustration of how heat does work. An imaginary object is divided into two parts (*a*), and then heated on one side only (*b*). The increased molecular motion on the heated side exerts pressure against the barrier, causing it to move and resulting in work.

In the real world, the "object" in our abstraction is typically water and the barrier the blades of the turbine of an electric generator. Heat makes the molecules of water push against each other until some burst free from the confines of the liquid as steam, which in turn pushes against the turbine's blades, making them turn to generate electricity. In our abstraction and also in the real world, it is not heat per se that does work, but rather a temperature difference between two places. If we were to heat the entire planet uniformly, no work could be obtained from the energy so expended. This, in a nutshell, is why dispersed energy is low-quality energy: it cannot do work. Now we can see why sunlight is a low-quality energy; it is dispersed all over the face of the globe and cannot be concentrated easily to produce the temperature difference required to do work.

Likewise, when coal or oil is burned, none of the energy is lost (first law of thermodynamics), but it is all dispersed, with a resultant increase in entropy (second law of thermodynamics). Energy so dispersed cannot do work unless it is reconcentrated; but the second law dictates that energy cannot be reconcentrated without an even greater expenditure of some other energy. Now we can understand why, delightful as the prospect may seem, energy cannot be recycled. Industrialism is caught inescapably in the iron grip of entropy. By accelerating the conversion of chemical and nuclear energy to heat, industrial societies also quicken the pace of the universe's monotonic march toward randomness.

Thermal Efficiency

When heat is converted to work, some heat is always "lost" (that is, dispersed) in accord with the second law. The efficiency of any machine can be measured by the proportion of heat so lost. Real-life efficiencies are startlingly low. A field of golden grain, for example, converts only 1% of the incident solar radiation to plant biomass,[2] as we shall discuss in greater depth later. An automobile is never more than 25% efficient.[3] That is, for each 4 gallons of gasoline burned, only 1 gallon is converted to useful work; the rest is discharged to the atmosphere as "waste" heat. Power plants that are fired with fossil fuels reach 40% efficiency. That is, for each 100 pounds of coal burned, 40 pounds are converted to the useful work of turning turbines. The other 60 pounds are discharged into the atmosphere as "waste" heat. Owing to unavoidable engineering limitations, nuclear power plants are usually less efficient than fossil-fuel plants. Nuclear plants seldom exceed 30% efficiency; hence, for a given amount of useful work nuclear plants generate substantially greater waste heat or thermal pollution. We shall return later to this point.

Where Does the Earth Get Its Energy?

There are 2 billion known galaxies in the universe. One of them, known to certain of its occupants as the Milky Way, has halfway toward its periphery an inauspicious star of medium size and magnitude, which like other stars constantly beams its energy into space. A tiny fraction of this radiant energy is intercepted by the star's third planet. The star is our sun; the planet, earth. The radiant solar energy intercepted by the earth is a tiny fraction of the sun's output, but enormous in human perspective. Solar energy flux at the earth's surface is thousands of times the summed energy budget of all industrial nations.

Besides the sun, there are two other substantial sources of energy on planet earth, geothermal and tidal. Scientists think that the earth was once a whirling sphere of hot gas which gradually coalesced into a ball of molten liquid. As the earth cooled, its crust became solid rock, converted eventually by wind, rain, and temperature fluctuations to soil. But beneath its comparatively placid surface the earth remains at heart the molten fury that it once was, as evidenced by infrequent volcanic eruptions. The heat stored in the earth's interior is still conducted to the surface at a rate several times the energy budget of industrial nations, and this heat is called geothermal energy. Tides, the rise and fall of large bodies of water caused by the gravitational attraction of the sun and moon, also contribute to the earth's energy budget.

Of these three forms of energy on earth, solar energy contributes $174,000 \times 10^{12}$ watts,* 5,000 times greater than the contribution of geothermal energy (32×10^{12} watts) and 58,000 times greater than the tidal input (3×10^{12} watts).[4] To understand the energy balance of our planet, then, clearly requires that we focus on the overwhelming contribution of the sun.

What Happens to Solar Energy on Earth?

Reflection

Imagine walking toward a white building on a hot day. Even before the building is reached one can feel heat radiating from it. The same thing happens on a celestial scale; fully 30% of incident solar radiation on earth bounces straight back into space, owing to the earth's reflectance or albedo. The whiter a body, the more sunlight it reflects, which may partially explain why desert-dwelling people wrap themselves in white robes and turbans. Sunlight bounces off anything that is white, including clouds, snow and ice cover, and airborne particles. As we shall see, increasing the albedo by atmospheric particle pollution is one of the ways that industrialism contributes to climate modification.

Absorption

If we approach our metaphorical white building and touch it, it feels warm. That's because in addition to reflecting sunlight, the building absorbs the sun's rays. The darker a body, the more sunlight it absorbs. Fully 47% of the sunlight that strikes the earth is directly absorbed by land masses and converted to warmth that helps make our planet such an

---

* A watt is a measure of energy flow per unit time. The human body generates about 100 watts.

agreeable place to live. Another 23% of incident solar radiation is absorbed in the seas where it evaporates ocean water into clouds and thus powers the hydrologic or water cycle. A fraction of this solar energy is retrieved by humans, for when rainwater ends up in high reservoirs, it has potential energy that can be converted to mechanical power or hydroelectricity.

The Climate

If we seem to have accounted for 100% of the solar radiation incident on earth, it is only because we have been inexact. A tiny fraction, 0.2%, maintains the oceans and atmosphere in circulation, contributing to the climate. The winds are driven by nature's grandest heat engine, the temperature difference between the poles and the equator. When sunlight strikes the earth, it beats directly down on the equator but glances off the poles at an angle. Moreover, equatorial regions are darkly colored as a consequence of lush tropical forests and hence they absorb heat, while the poles are white and thus reflect heat. As a result of these simple physical forces the equatorial regions are warm, the poles frigid. Recall that when air is heated, as above a candle, it rises. Air above the equator follows this rule, and towers miles into the tropical sky. At the poles, in contrast, air is cold and dense and sinks down into a compact, low mass. When the hot air at the equator rises, it creates a relative vacuum. Nature, of course, abhors a vacuum, and cold air from the poles therefore slips into the place of the risen hot air. Meanwhile, the high atmosphere above the equator slides down a long inclined plane toward the poles. Hence, the atmosphere is in constant circulation, from poles to equator, up and back again to poles, in yet another of nature's magnificent cycles (Figure 1.4a). Now give the globe beneath the atmosphere a spin so that its movements too are cyclic, and the tropical air rising at the equators spirals toward the poles to create the basic wind patterns of planet earth (Figure 1.4b). Humans have tapped this solar energy throughout recorded history with windmills and sailing ships.

The boundary at which the hot air of the tropics meets cold blasts of air from the poles is called the polar front. Along this front, masses of hot and cold air dance furiously about each other in gigantic spiraling whirlwinds called the circumpolar vortexes. Another process unfolds along the polar front, a process familiar to anyone who has blown hot breath against a cold window pane. The hot, moist tropical air meets the cold polar air and contracts, releasing its moisture in the form of clouds and rain. The clouds outline the shape and mark the position of the circumpolar vor-

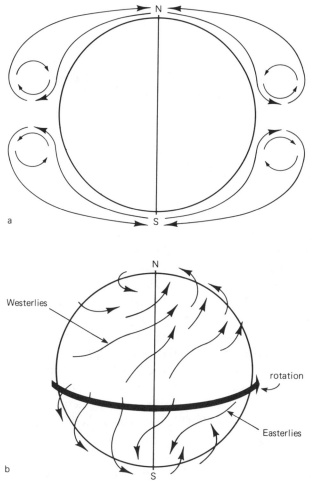

1.4 Generation of the winds by nature's grandest heat engine, the temperature difference between the poles. If the earth were stationary, the winds would look like the pattern in *a* (see text). Since the earth rotates from west to east, the actual wind patterns occur as shown in *b*.

texes, as is familiar to television viewers who have seen satellite weather pictures of storms on earth.

Photosynthesis

We shall have more to say about climate, but in the meantime we have not yet accounted for the small fraction of incident solar energy—a mere two one-thousandths of the total—that is the most directly critical to industrial societies. It is this fraction that is captured by green plants in the process of photosynthesis, to be converted by chemistry and time to the fossil fuels on which industrialism has depended. To put it differently, of the almost insignificant fraction of the sun's energy that is intercepted by planet earth, an almost insignificant proportion of that powers photosynthesis, which is in turn the energetic basis of life on earth.

Entropy and Life

We have observed that energy is never lost when it changes from one form to another (the first law of thermodynamics), but it is always degraded in the sense that it is dispersed, with the inevitable increase in entropy (the second law). At first sight the laws of thermodynamics might seem inconsistent with the existence of the most highly organized phenomenon in the known universe, life. Most cells in our body are far smaller than the head of a pin and yet so wondrously complex that until a few decades ago biologists believed that each cell must be invested with its own share of a "vital substance" that was inaccessible to scientific analysis. Molecular biologists replaced this philosophy of "vitalism" with that of "mechanism" by showing that each cell is in reality a finely tuned miniature factory, possessing all the machinery needed to process delivered foodstuffs, export unnecessary waste products, and replicate and rebuild itself. The wonder of life is hardly dispelled by these discoveries: imagine, if you will, the task of designing and constructing such a miniature factory, and then assembling several into a bee, a flower, or a human brain! The most ambitious engineering scheme ever conceived by the human mind pales in comparison.

How could the marvelous organization implicit in living matter exist in the face of a universal tendency toward randomness? The key to the puzzle lies in the definition of entropy. We have said that within a closed system, entropy *on the average* increases. Within restricted regions of a closed system, however, entropy may decrease, but only if such decrease is more than compensated by a disproportionate increase elsewhere in the closed system. Fortunately for life on earth, our planet is not a closed system; as we have seen, it receives an abundant if diffuse energy input from the sun. Evolution has chanced on a way to upgrade this low-quality solar radiation into highly organized living matter or biomass on earth, but only at the cost of a disproportionate increase in entropy somewhere else in the system, namely, the sun itself. The mass of the sun is declining measurably. If our sun follows the known pattern of other stars, countless eons from now it will expand precipitously into a red giant and then, in a final blaze of glory, collapse into a white dwarf, leaving in its darkened wake a frozen earthly record of life, a veritable treasure chest for intergalactic anthropologists of the distant future.

NOTES

1. G. M. Crawley, *Energy* (New York: Macmillan, 1975), p. 44.
2. E. J. Kormondy, *Concepts of Ecology*, 2nd ed. (Englewood Cliffs, N.J.: Prentice-Hall, 1976).
3. H. E. White, *Modern College Physics*, 3rd ed. (New York: D. Van Nostrand, 1956), p. 318.
4. All specific values reported in this chapter for natural energy budgets of our planet are taken from M. K. Hubbert, *U.S. Energy Resources, A Review as of 1972*, Document #93–40 (92–75) (Washington, D.C.: U.S. Government Printing Office, 1974).

ADDITIONAL REFERENCES

The "energy crisis"of the early 1970s spawned several books on the subject, and stimulated countless government reports and studies, such as Note 4 cited above. One of the best books I have seen for a general development of energy principles and sources for a nonscientific audience is Crawley's book *Energy* (cited above in Note 1). Readers interested in the ecological implications of energy for life will find endless food for thought in the brilliant little book by the Odums, which inspired many ideas in *The Seventh Year* [H. T. Odum and E. C. Odum, *Energy Basis for Man and Nature* (San Francisco: McGraw-Hill, 1976)].

The best sources for up-to-date appraisals of developing energy sources and problems are publications of the American Association for the Advancement of Science (AAAS). These include their weekly magazine *Science* (e.g., several articles in the issues of 10 Feb. 1978 [volume 199, no. 4329], and 14 April 1978 [volume 200, no. 4338], as well as occasional, timely special publications such as P. H. Abelson, ed., *Energy: Use, Conservation and Supply* (Washington, D.C.: AAAS, 1974), and A. L. Hammond, W. D. Metz, and T. H. Maugh III, *Energy and the Future* (Washington, D.C.: AAAS, 1973). Future chapters will draw heavily from specific articles in these sources.

By way of a general bibliography on energy, I have found the following sources helpful as background or in preparing *The Seventh Year*:

Davis, W. K. (ed.). *U.S. Energy Prospects: An Engineering Viewpoint*. Washington, D.C.: National Academy of Sciences, 1974.

Freeman, S. D. *A Time to Choose: America's Energy Future*. Cambridge, Mass.:
    Ballinger, 1974.
        This is the final summary report by the Energy Policy Project of the
    Ford Foundation. The full report occupies some 20+ individual books, pub-
    lished also by Ballinger, some of which are cited in future chapters.
Freeman, S. D. *Energy: The New Era*. New York: Vintage, 1974.
Lovins, A. B. *World Energy Strategies: Facts, Issues and Options*. San Francisco:
    Friends of the Earth, 1973.
Lovins, A. B., and Price, J. H. *Non-Nuclear Futures*. San Francisco: Friends of
    the Earth, 1975.
Kenward, M. *Potential Energy: An Analysis of World Energy Technology*. New
    York: Cambridge, 1976.

Chapter 2
THE EVOLUTION
OF INDUSTRIALISM

Since the dawn of human history our activities have been governed by the laws of thermodynamics introduced in the preceding chapter. In the present chapter we will see how the evolution of industrialism can be understood in terms of these laws. We will discover that the onset of industrialism, a recent event in the perspective of human history, represents an extension of age-old energy patterns to a new stage of development in which certain of nature's cycles are dramatically accelerated. We will also see how large-scale industrialism deviates from certain laws of nature. It is because of these deviations from natural law that the affluent way of life so characteristic of advanced industrial nations cannot be sustained indefinitely.

How Is Energy Made Available to Living Forms?

As we have seen, solar energy is captured, or "fixed," by photosynthesis. It is possible to discover how much energy is fixed by green plants simply by weighing them and determining by a simple calculation how much chemical energy is contained in the form of natural sugars. The efficiency of a green plant is expressed as the proportion of incident solar energy that is actually fixed as chemical energy. For a single plant, about 1% of the incident solar radiation is captured. Of this amount, 50%–75% is used by the plant to carry on its life activities, including growth, development, and metabolism. The remainder is converted by the plant to stored chemical energy. Overall, then, the "gross" production efficiency of a field of plants is in the neighborhood of 1%, the "net" efficiency 0.5%–0.75%.[1] In other words, if 2,000 calories* of solar energy fall on a field, only 20 calories end up stored as living matter, or biomass, in the form of plants. This small fraction of energy has been upgraded, however, because diffuse sunlight has been concentrated in the form of chemical energy. Ecologists call green plants the primary producers of nature, because only they possess within their cells the biochemical equipment to "fix" solar energy. In reality, of course, energy is not produced by green plants—it is merely converted from one form, sunlight, to another, chemical energy, in accord with the laws of thermodynamics. This "fixed" chemical energy is in turn the energetic basis of all other life on earth.

What Happens to Green Plants?

After green plants fix solar energy, they either die and decay, are buried to become fossil fuels of the future, or they are eaten by animals that can digest them, the herbivores (antelopes, deer, elephants, kangaroos, etc.). We have seen that every time energy is converted from one form to another some of it is dissipated in the form of heat. The same law of thermodynamics applies to energy conversions by plants and animals. The conversion efficiency of herbivores can be determined simply by weighing them and finding out what fraction of incoming chemical energy (food) was actually incorporated into their bodies as animal biomass. It turns out to be surprisingly low, only 10%. As plant biomass is transformed to animal biomass, however, energy is again upgraded, or concentrated, for the molecules of animals (notably fats) are richer in energy than those of plants (carbohydrates).† This is partly why herbivores such as cows need to graze continuously to obtain the chemical energy they need, while carnivores like wolves can do with a meal only once every few days.

---

*A calorie is a measure of heat energy, and is defined by physicists as the amount of heat required to raise the temperature of 1 gram of water by 1 degree Centigrade.[2]

† A gram of protein or carbohydrate yields about 4,000 calories, while a gram of fat yields about 9,000 calories. Animal tissue is more energetically "expensive" than plant because animals contain more protein and fat than plants, which are largely carbohydrate.[3]

What Happens to Herbivores?

Ecologists say that green plants and herbivores represent two different energy levels, or *trophic* levels, and that this relationship is the first link in the *food chain*. The second link in the food chain is completed when carnivores, or meat-eaters, consume the herbivores. Energy is dissipated in this step too: only 10% of the chemical energy stored in the tissues of herbivores is incorporated into the tissues of carnivores. The other 90% is lost in a variety of ways, but eventually dissipated as heat. No energy is upgraded in the conversion of herbivores to carnivores, however, since the tissues of carnivores are made up of the same energy-rich molecules as those of herbivores. We can see why, in energetic terms, it is more efficient for people to eat vegetables than meat; by doing so we bypass one trophic level in the food chain and thus avoid the attendant dissipation of heat energy which would occur according to the second law of thermodynamics. The laws of thermodynamics thus explain why many more human beings can be fed from grain directly than if the grain is first converted to cows which are then eaten.

What Happens to Carnivores?

The term "food chain" is somewhat misleading, for in terms of the disposition of nutrients it is actually a cycle. Recall that in the process of photosynthesis plants produce not only sugar, but also oxygen. Herbivores eat the sugar and breathe the oxygen. Within the tissues of animals the oxidation reaction takes place, liberating energy for animal life. Animals then complete the exchange with plants by expiring carbon dioxide that in turn supplies photosynthesis. When animals die they are either eaten by scavengers or their tissues are reduced to carbon dioxide and other basic elements by bacteria. These elements enter the soil to become incorporated again into green plants. In this way nature reuses the same resources again and again in natural cycles that are driven by the constantly renewed energy from the sun. Energy is dissipated through a "chain," but nutrients themselves are constantly recycled. Moreover, and as we shall examine soon, organisms in the food chain return energy to the chain, with the effect of accelerating the steps of the chain. The chain is thus closed and hence takes the form of a cycle. Life itself, like the movements of the planets in the heavens and water on earth, is cyclic. The same atoms that comprise our flesh and bones may once have been contained in the bodies of our ancestors, and will presumably be contained in the bodies of our children's grandchildren.

Ecology Equals Thermodynamics

In biology, as in physics, the laws of thermodynamics thus reign supreme. Energy is never created or destroyed: it is merely converted from one form to another in the food cycle (Figure 2.1). Each time such a conversion is made, however, substantial energy is dissipated in heat that cannot do work. That is, entropy increases, in accord with the second law. The energy that is not dissipated is upgraded in the sense that it is concentrated. The energy that is dissipated is absorbed by the earth or radiated into space, to be replaced by the continuous flow of solar energy to the earth.

Cultural Evolution and Natural Law

Like other animals from which our species has evolved, the human organism has always participated in the upgrading of energy.[4] But unlike most other organisms, human beings form societies, which in turn furnish the basis of culture. In the process of cultural evolution, humankind has carried the process of energy upgrading to extremes that are unknown elsewhere in nature. Such extremes are made possible by humanity's increasing reliance on vast quantities of *stored* energy and resources. As we have seen, animals in nature also depend on energy and resources that are stored in other organisms; but their needs are modest in magnitude, and their supply is replenished continuously and on a rapid time scale. In contrast, the industrial civilization of which we are a part now depends for its existence on vast quantities of energy and resources, stored largely in fossil fuels and in rich mineral deposits that require millennia to accumulate. Unlike other animals, we are now consuming our energy and resource base much faster than it can be replaced by nature.

In future chapters I will develop the thesis that this uniquely human deviation from natural law is self-reinforcing; and I will argue that in this

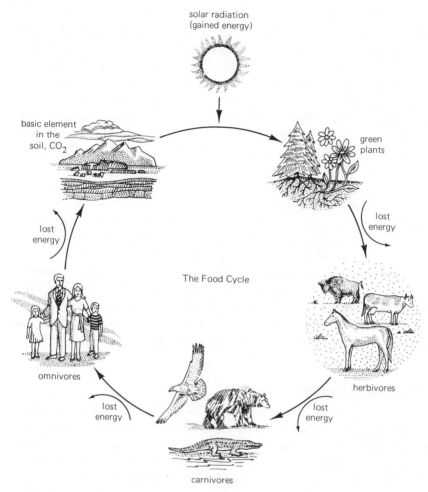

solar radiation
(gained energy)

basic element
in the
soil, CO_2

green
plants

lost
energy

The Food Cycle

lost
energy

omnivores

herbivores

lost
energy

lost
energy

carnivores

2.1 The food cycle, by which nature converts energy from one form to another. Basic elements in the soil are combined with carbon dioxide (CO_2) by photosynthesis (see Figure 1.2) to produce plants, which in turn serve as the energetic basis of all other life forms. The entire cycle is driven by energy from the sun. Energy is lost at each conversion in the cycle, in accord with the second law of thermodynamics.

deviation the seeds of the decline of industrialism are already visibly sown. But first let us seek the broad perspective of time. Let us step back in history, to the dawn of human culture, and try to identify the roots of our present patterns.[5]

Hunter/Gatherer Societies

Anthropologists think that the most "primitive" preindustrial cultures were hunter/gatherer societies. In these societies humans occupied a relatively simple position at the upper end of natural food chains that were driven exclusively by solar energy. Roots, berries, wild fruits and nuts, together with wild game, comprised the diets of these peoples. Essential shelter and clothing needs were likewise met directly by nature.

We are accustomed to thinking of hunter/gatherer societies as "backward," lacking a significant culture and reduced to grubbing in the ground for a barely adequate ration of roots. Modern anthropological studies, however, paint a strikingly different picture. The !Kung Bushmen of Africa, for example, live in hunter/gatherer societies that until recently had changed little since the beginning of history.[6] The people are intelligent, extraordinarily healthy and long-lived, nonviolent, and nonsexist. They instruct their children in cooperation rather than competition. Their food is easily provided by two or three days' work per week; with the remaining time they have developed a rich and varied culture, involving ritual, elaborate dance and music, and religion. Hunter/gatherer societies were apparently primitive only in the narrow sense that they made no deliberate attempt to accelerate the upgrading of solar energy. One wonders whether these pristine people had an instinctive awareness of the human position in nature's cycles. They were true solar societies, and it may be no coincidence that they often worshipped the sun.

Shifting Agriculture

In parallel with hunter/gatherer societies there evolved a new and energetically more "advanced" form of human culture in which natural food sources were supplemented by slash-and-burn agriculture. These societies, usually located in the great tropical rain forests of the world, cleared a small area of forest in which soil nutrients had accumulated for several years. The cut trees and brush were then burned and the ashes distributed over the cleared area to provide the soil with additional nutrients. Patches so cleared could be used to grow simple crops for two or three years. By the end of this period, however, the intensive tropical ecosystem reestablished itself. The forest reclaimed the cleared land and the society moved on to the next site, revolving through several such sites and back again in a period of 10–20 years. Small-scale domestication of animals, especially wild pigs, was sometimes practiced by such societies. This pattern of shifting agriculture discouraged the construction of stationary civilizations, but judging from the few such extant tribes did not prevent the evolution of a complex society that was finely tuned to the natural cycles in which it was embedded.

In terms of energetics, slash-and-burn agricultural societies represented a significant advance over hunter/gatherer societies. Here for the first time humans made use of stored energy beyond that contained in the immediate food chain, in the form of nutrients built up in the soils and burned vegetation. Here also for the first time high-quality energy in the form of human labor was deliberately applied to accelerate the upgrading of low-quality solar energy into food by a simple agriculture.

As a result of accelerated energy fixation and utilization, shifting agricultural societies laid the basis for an increase in the size of the human population. In thermodynamic terms, accelerating the upgrading of solar energy increased the amount of high-quality energy contained in the form of human biomass. Such a process is in principle self-reinforcing, for the more human labor available, the more food that can be grown to support the growth of yet more human beings. Such a self-reinforcing process is termed positive feedback. In practice, however, the limited efficiency of the slash-and-burn agriculture severely constrained population growth; the stage was not yet fully set for the positive feedback cycle of runaway population growth.

Stable Agriculture

As human populations slowly grew, they gradually depleted the wild populations of plants and animals on which they depended for existence. Several big-game species were hunted to extinction in Europe and America by 9,000 B.C.[7] In ecologic terms, the reservoir of upgraded solar energy that was stored in the biomass of natural populations of plants and animals was drained by human activity. Anthropologists now recognize the possibility that the ecological pressures of the growing human population made domestication of plants and animals a necessity for survival in many regions of the planet.[8] As we have seen, the intense climate and soil conditions of tropical regions are not well suited to a geographically stable agriculture. In contrast, when land in the higher latitudes is cleared for crops it remains relatively open and can, with proper farming practice, be planted year after year. The possibility of geographic permanence permitted, and ecologic pressure apparently required, the development of stable agricultural societies. In these societies human labor could be systematically directed toward the upgrading of solar energy into domestic plants and animals. The animals included not only those that could be used for food, but also beasts of burden.

The efficiency of production associated with a stable agriculture heralded a new era of human history. For the first time food surpluses were available, freeing the labors of a fraction of the populace for pursuits other than food production. Metal was discovered and fashioned into tools. Great cities were built and monuments such as the pyramids erected as lasting records of the early human civilizations. Writing was developed as a means of keeping the records essential to the administration of an increasingly complex and specialized civilization. Certain members of society were freed from the process of food production to seek knowledge, which may be seen as a highly upgraded form of energy that depends for its production and transmission on an excess of lower quality energy, stored in the forms of food, shelter, and the physical trappings of learning. The earliest knowledge was of nature, knowledge that was in turn directed toward perfecting agricultural practices and thus accelerating the further upgrading of solar energy. Finally, the surplus energy generated by a stable agricultural society permitted the assembly of gigantic, well-equipped armies, whose unconscious purpose was the preservation of upgraded energy (defense) and its concentration in the hands of those who commanded the generals' allegiance (offense). Wood, a direct product of photosynthesis, provided fuel for smelting iron and timbers for ships of commerce and war. Towering virgin cedars, the stored solar energy of millennia that graced the shores of Lebanon, were harvested beyond their capacity to regenerate. Thus was the ascent of the human species propelled in part by the deforestation of the Cradle of Civilization.

We see in stable agricultural societies the repetition of the now-familiar energy patterns. Upgrading of solar energy through agriculture was accelerated by reliance on energy stored in trees in the form of wood; and the upgraded energy itself, stored in surplus food, buildings, knowledge, larger populations and armies, was "fed back" to the food cycle, i.e., utilized to accelerate even further the upgrading of solar energy. Human civilization stood on the threshold of industrialism.

The Industrial Age

The cultural and technological evolution sketched above occurred over thousands of years. Because the growth of technology was slow compared to human reproductive capacities, the expansion of the human population could easily keep pace with the expansion of energy conversion. Thus the per capita energy conversion (i.e., the energy consumption per person) probably remained essentially unchanged from the beginning of the Stone Age to a thousand years ago.[9] This circumstance was dramatically changed, however, by the discovery nine centuries ago that certain black "rocks" littering the beaches of Britain burned extraordinarily well. These "sea coals," fragments of coal eroded from exposed seams by the pounding action of the waves, were at first little more than a convenient supplement to wood for heating England's coastal households. But the British government, faced in the middle of the millennium with a severe shortage of oak for the construction of its fleet, was prodded by the Admiralty to husband England's remaining hardwood forests. It did not take the enterprising British long to realize that coal was a suitable substitute for wood for smelting metal and other manufacturing purposes. Miners followed exposed seams of coal into the earth, and the first coal pits were dug. Neither did it take the British long to learn that pits dug into the earth typically fill with water. The water was at first laboriously removed by hand, and later by elaborate belts of buckets coupled by shafts and pulleys to treadmills on which oxen were made to walk. The pattern thus repeats: high-quality animal energy was "fed back" to extract coal, which provided energy that ultimately made possible the further upgrading of solar energy. The exploitation of the immense stores of energy contained in fossil fuels began in earnest.

Coal is a bulky fuel that cannot easily be transported over great distances; hence it had to be used initially near its source. It is no accident that six of the seven largest metropolitan regions in modern England are built upon ancient coal seams now in various stages of exhaustion. The development of electricity in the early 1800s, however, freed industrialism from this geographical constraint, for unlike coal, electricity is easy to transport over long distances. The stage was set for the accelerated expansion of industrialism in the 20th century.

Coal initially fueled the Industrial Revolution. It was used first to heat homes and smelt iron ore and later to fuel steam engines and generate electricity. In particular, the early wedding of the coal and iron industries established a synergic relationship that was to form the structural basis of modern industrialism. In the mid-1800s, however, well after industrialism had spread to the European continent, an event of immense significance occurred in Poland. A group of enterprising farmers, convinced that the black, sticky liquid that oozed from the earth in certain locales could be put to some useful purpose, persuaded a pharmacist to try to distill from it their cherished vodka.[10] Instead the pharmacist obtained kerosene, and in the process discovered how to refine oil. Not long after, a retired railroad conductor with an inclination for mechanical tinkering drilled the first oil well near Titusville, Pennsylvania, and struck a gusher at a mere 69 feet.[11] Entire cities of derricks arose overnight (Figure 2.2), and the earliest refineries were constructed to process the "black gold." Industrialism in America had entered a century-long period of unremitting growth.

2.2   An early oilfield at Signal Hill, California. (courtesy Shell Oil Company)

The Lessons of Ecocultural History

The evolution of human culture from hunter/gatherer societies to industrialism is characterized by the repetition of several tightly interwoven energy patterns. An explicit recognition of these patterns will help us to appreciate what is new and different about industrialism, and why it is ultimately a self-limiting process.

Energy Upgrading

First, we have seen that each human culture, from the most "primitive" to the most "advanced," has busily if unconsciously engaged in upgrading solar energy. Hunter/gatherer societies upgraded solar energy with a minimum input of human labor simply by occupying the upper end of the food chain. Early agricultural societies accelerated the food cycle by more intensive application of human labor and the harnessing of animal energy to aid in the cultivation of food. Finally, industrial societies such as our own manufacture tractors and fertilizers that vastly increase per capita agricultural productivity, and also engage in agricultural research that produces knowledge and accelerates the food chain even further. All of these cases share a critical common feature, the feedback of high-quality energy to the chain of events by which solar energy is upgraded into plants and animals[12] (Figure 2.3). Industrial societies are unique in their reliance on massive amounts of stored energy and resources to accelerate the food chain, as we see next.

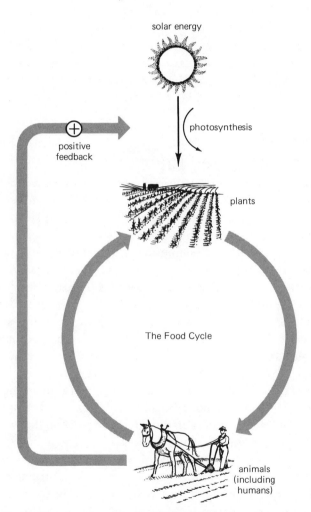

2.3   Positive feedback acceleration of the food cycle. Human cultures, both prehistoric and modern, accelerate the food cycle by using high-quality energy (human and animal labor) and, in the case of industrial societies, fossil fuels.

Energy Subsidization

The second energy pattern that is repeated in the evolution of human culture is that each cultural transformation to a new energy source has been "subsidized" by the energy that fueled the preceding culture. Wild plants and animals subsidized the formation of stable agricultural societies. Wood and animal power were used to mine the first coal in England; only later was coal itself used to operate the mining machinery. In the U.S., wood was used in the early 19th century to subsidize the development of fossil fuels, with the consequent deforestation of vast regions of the nation. And now the age-old pattern is repeating once again. Fossil fuels—coal, oil, and natural gas—are being used to support the research and the construction of facilities needed to harness a new source of power, nuclear energy. That is, fossil fuels are subsidizing the development of nuclear power.

Reliance on Stored Energy

Finally, a third energy pattern is repeated in the evolution of human culture, namely, the dependence on stored energy. Animals in nature also depend on the energy stored in other organisms (plants and animals) for their survival. But the ingenuity of our species has permitted elaboration on this theme, to the degree that a qualitatively new phenomenon has emerged. Each transition to a more "advanced" human culture was characterized by the exploitation of a more concentrated, or higher quality, energy form. The solar energy on which the most "primitive" peoples depended was most diffuse; wood is a concentrated form of solar energy; coal is more concentrated yet, in that the heat capacity per unit weight is greater; and finally, humanity is drawing on the most concentrated form of stored energy that is known—nuclear energy.

As we shall develop, reliance on the immense quantities of energy stored in the chemical bonds of fossil fuels has released, temporarily, the natural ecological constraints on the human population that were imposed by dependence on the constantly renewed but low-quality energy of the sun. Society's use of concentrated, stored energy instead of constantly replenished solar energy is analogous to making a construction loan from a bank. The capacity for doing work that is implicit in the large amount of capital "stored" by the bank is of course much greater than that which is possible from a steady but small income. The catch, of course, is that the loan is eventually exhausted—and then it must be repaid.

The Role
of Positive Feedback

We see that within limits even history is cyclic, in the sense that cultural evolution is characterized by the repetition of the same basic energy pattern. And yet there is clearly something unique about modern industrial culture, something that makes it a qualitatively new phenomenon in comparison to preindustrial cultures. What makes industrialism unique? The answer lies in the nature of positive feedback. We have seen that the upgrading of solar energy to plants by the active feedback of energy is a self-reinforcing, or *positive-feedback*, process, in that more plants (low-quality energy) yield more food for more people (high-quality energy), increasing the amount of human labor available to "feed back" to the cultivation of plants. Prior to industrialism, the positive feedback process of energy upgrading by human activity (Figure 2.3) did not yield uncontrolled, runaway population growth. With access to increasingly concentrated stored energy, however, industrial societies have broken free of these constraints. Natural limits have temporarily been transcended by the unprecedented conversion of vast quantities of stored energy. Per capita energy conversion has increased from an estimated 500 watts per person in preindustrial societies to 2,000 watts per person at present, and is growing at a rate of 5.7%, three times faster than world population growth of 1.9%.[13]

Why was the positive-feedback process of energy upgrading stable in preindustrial societies but uncontrolled now, resulting in runaway growth? The answer is best illustrated first in a simple abstraction which we shall then apply to industrialism. Let us imagine a simple positive-feedback process in which process a causes process b which causes a (Figure 2.4). Let us assume further that both a and b can assume specific numerical values which express the *rate* of the corresponding processes. The operation of this loop is defined by what engineers call its *gain*, i.e., the ratio of the output to the input at any place in the loop. Suppose first that the gain is 0.9; that is, an input of 10 at any place in the loop will cause an output of 9. This being the case, an initial value of 10 at a will become 9 at b, 8.1 at a, 7.29 at b, etc. In other words, even though the overall dynamic is self-reinforcing, with each iteration the process gradually "runs down." If we graph the activity of the loop over time, we see that the decline takes the form of what mathematicians call an exponentially declining curve (Figure 2.5a).

Now consider what happens when the gain of our positive-feedback

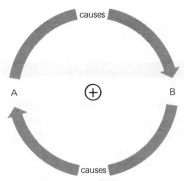

2.4 Schematic illustration of a simple positive-feedback loop, in which two processes are linked by reciprocal causation.

loop is changed to 1.1. An *a* value of 10 causes a *b* value of 11 which causes an *a* value of 12.1, etc. If we graph the activity of this loop over time, we see that the curve increases exponentially (Figure 2.5b). Using the same logic, the reader may verify that when the gain is set exactly at 1.0, the output of the positive-feedback loop remains stable over time.

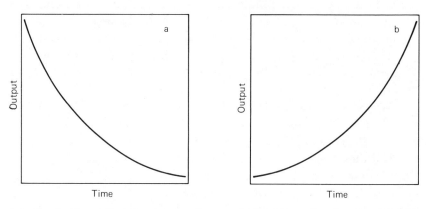

2.5 Graphs illustrating the output of any positive-feedback loop when the gain is less (*a*) or more (*b*) than 1.0. When the gain is exactly 1.0, the output over time is constant, and the graph takes the form of a straight horizontal line.

Our abstraction thus illustrates that the behavior of a positive-feedback process is governed by its gain. When the gain exceeds unity (1.0), the positive feedback is bound to yield runaway exponential growth. In the case of modern industrialism this is precisely what has happened: the gain of the age-old positive-feedback process by which solar energy is upgraded (Figure 2.3) has at last exceeded unity, yielding runaway exponential growth.

What does it mean in practical terms when the positive-feedback cycle of energy upgrading exceeds unity? Reduced to fundamentals, it means that growth can take place; the more the gain exceeds 1.0, the greater the rate of exponential growth. Let us illustrate with the example of population. Population growth is self-reinforcing, in the sense that people produce people. When the gain of this process is less than 1.0, population automatically declines; when it is exactly 1.0, population is stable; but when the gain exceeds 1.0, then by definition enough solar energy can be upgraded to food so that one couple can reproduce and raise to sexual maturity more than their replacement value of two offspring. Such a condition obtains when large quantities of high-quality stored energy is fed back to the food chain. Thus, it is ultimately the use of concentrated, stored energy, rather than constantly renewed solar energy, that fuses the "population bomb." In this sense, abundant energy is responsible for the explosion in human numbers that has occurred during the Industrial Age (see Chapter11).

But it is not only the human population that has grown exponentially; it is everything—energy conversion (the driving force), resource utilization, knowledge, technology, material wealth, etc. As we shall illustrate in future pages, these variables also grow by positive-feedback cycles that are linked to energy conversion. Indeed, everything in industrial societies, including change, is speeded up by "runaway" positive feedback, contributing to what Toffler has termed "future shock."[14]

Perhaps the future will shock us less if we understand the forces that shape it. As we have seen in this chapter, and as has been clearly appreciated by others,[15] stored energy is the driving force of the industrial civilization that now inhabits the earth. As the present constrains the future, so do present patterns of energy conversion influence the future of industrialism. If we are to begin to understand the future of our civilization, we should look next to how industrial society uses energy in the present. Then we must examine the sources of this energy and try to assess the likelihood of their availability in the future. It is to these topics that we now turn.

NOTES

1. E. J. Kormondy, *Concepts of Ecology*, 2nd ed. (Englewood Cliffs, N. J.: Prentice-Hall, 1976).
2. H. E. White, *Modern College Physics*, 3rd ed. (New York: D. Van Nostrand, 1956), p. 272.

3. Kormondy, *Concepts of Ecology*.

4. H. T. Odum, and C. Odum, *Energy Basis for Man and Nature* (New York: McGraw-Hill, 1976).

5. O. D. Duncan, "Social organization and the ecosystem," in *Handbook of Modern Sociology*, edited by R. E. L. Faris (Chicago: Rand McNally, 1964), pp. 127–159; W. Goldschmidt, *Man's Way: A Preface to the Understanding of Human Society* (New York: Holt, Rinehart and Winston, 1959).

6. G. B. Kolata, "!Kung Hunter-gatherers: Feminism, Diet, and Birth Control," *Science* 185 (1974): 932–34.

7. M. Harner, "The Enigma of Astec Sacrifice," *Natural History* 86 (1977): 47–51.

8. M. N. Cohen, *The Food Crisis in Prehistory* (New Haven, Conn.: Yale University Press, 1977).

9. M. K. Hubbert, *U.S. Energy Resources, A Review as of 1972*, Document #93-40 (92-75) (Washington, D.C.: U.S. Government Printing Office, 1974).

10. N. Grove, "Oil, the Dwindling Treasure," *National Geographic* 145 (no. 6): 792–825.

11. A. Toffler, *Future Shock* (New York: Bantam, 1970).

12. Kormondy, *Concepts of Ecology*, pp. 41–42.

13. Hubbert, *U.S. Energy Resources*.

14. Toffler, *Future Shock*.

15. Odum and Odum, *Energy Basis for Man and Nature*.

Chapter 3
THE ENERGY BASIS
OF INDUSTRIALISM

Where do industrial nations obtain the energy they need to keep their factories humming? Initially, as we saw, the Industrial Revolution was fueled by coal. England, the grandmother of the Industrial Age, still depends on coal for nearly half its energy needs (Table 3.1). But coal is a dirty fuel, awkward to handle and dangerous to mine. Its disadvantages, together with the accidents of geographic history that make one nation rich in oil and another in natural gas, have caused industrial nations to diversify somewhat in their reliance on fossil fuels. The major industrial nations still depend on fossil fuels for more than 95% of their energy needs, but the pattern varies from one nation to the next. Thus the United States led all nations in its dependence on natural gas, while Japan was equally committed to oil. Nuclear energy accounted for less than 5% of all energy conversion in 1970.

Table 3.1. FUELS OF SELECT INDUSTRIAL NATIONS IN 1970.[a]

Country	% solid fuel (mostly coal)	% liquid fuel (mostly oil)	% gaseous fuel (mostly natural gas)	% hydro-power	% nuclear power
United States	20.5	42.7	34.8	1.7	0.3
France	26.5	62.4	6.1	4.1	0.9
West Germany	39.3	53.3	5.8	1.0	0.6
Italy	7.6	77.9	9.8	4.2	0.5
England	48.0	44.2	5.0	0.2	2.6
Europe (OECD)	29.4	59.6	6.7	3.3	1.0
Japan	23.3	71.7	1.4	3.2	0.4

[a]Expressed as % of total energy needs met by the indicated source. From H. H. Landsberg, "Low-cost, Abundant Energy: Paradise Lost?" *Science* 184 (1974): 247–53.

Sources of Energy for U.S. Industrialism

The United States is responsible for about one-third of the world's total energy conversion. Thus it is an especially indicative nation for the future of industrialism. The sources of energy for the U.S. are illustrated in Figure 3.1. This figure is based on data compiled and projected into the future by the U.S. Department of the Interior; hence it provides a clue as to how the U.S. government has viewed its own energy future.

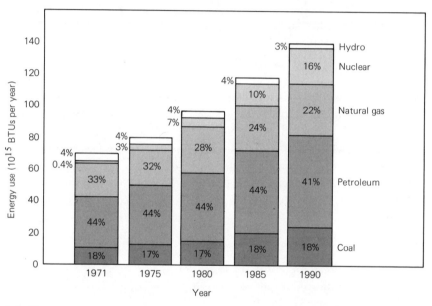

3.1 Energy consumption in the United States by source. BTUs = British Thermal Units, a measure of energy use; 1 BTU = 0.259 calories. 10^{15} BTUs ≈ 172×10^6 barrels of oil, 970×10^9 cubic feet of natural gas, or 41.7×10^6 tons of coal. (after G. A. Lincoln, "Energy Conservation," in *Energy: Use, Conservation and Supply*, edited by P. H. Abelson [Washington, D.C.: AAAS, 1974], Figure 2)

According to this view, energy conversion in the U.S. will more than double from 1971 to 1990, corresponding to an annual growth rate of 3.5%. The average annual worldwide rate of increase in energy use was 4.38% over the past three decades,[1] and, according to a recent U.N. report, 5% over the past 25 years.[2] Hence the U.S. government's projected estimate of 3.5% for its own annual growth rate is comparatively modest. In 1975, oil was the most important source of energy for U.S. industrialism, accounting for some 44% of the total energy budget. Oil was

followed by natural gas (32%) and coal (17%). Nuclear energy just sur-
passed firewood as a fuel source, accounting for 3% of the 1975 energy
budget. The most noteworthy feature of the government's projection to
1990 is the decreased use of natural gas (from 32% to 22% of the total),
compensated for by an increased dependence on nuclear energy (from 3%
to 16%).

Uses of Energy by Economic Sector

How does the U.S. employ the prodigous quantities of energy that it
uses? Relying again on the Department of the Interior for data, Figure 3.2
shows the actual pattern of energy use in 1971, and the projected pattern
of energy use through 1990, also based on a modest growth rate of 3.5%.
In 1971, roughly one-fourth of the energy budget was devoted to each of
the four major economic sectors, namely, transportation, residential/com-
mercial, industry, and electric utilities. Let us briefly examine each of
these in turn, in order to provide a background against which to assess our
future.

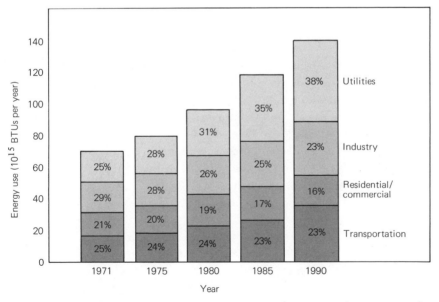

3.2   Energy consumption in the United States by consuming sector. (after G. A. Lincoln,
"Energy Consumption," in *Energy: Use, Conservation and Supply*, edited by P. H. Abelson
[Washington, D.C.: AAAS, 1974], Figure 1)

Transportation

The transport of materials is a critical function in every industrial society. Not only must people reach their workplace in order to keep the wheels of industry spinning; in addition, the products of industrialism must be transported to distant markets, and food and resources must be transported to the urban conglomerations where people and machines are concentrated. Fuels, wastes, and water must be transported to and from cities. In short, and as we shall develop in greater detail in later chapters, an effective system of transportation is crucial to the continued operation of industrial society.

In 1970, transportation in the U.S. was 96% dependent on the liquid fuels that are refined from petroleum. Travel by personal auto accounted for 55% of the total transportation energy, equivalent to 14% of the total U.S. energy budget. Trucks came second, using 21% of the transportation energy, and aircraft third, with 7.5%. In this century the abundance of energy has permitted a steady progression away from the slower, most energy-efficient modes of transportation (rail, waterways, mass transit) in favor of faster but less efficient modes (automobiles, trucks, and aircraft).[3] Projections to 1985 show an increased relative reliance on the less efficient means of transportation.

Residential and Commercial

More than half the energy consumed by this sector was used for space heating and cooling. The heavily populated northeastern and midwestern U.S. could not survive the frigid winters without substantial energy for heating, and the current population shift to the southern states depends critically on air conditioning during the hot and humid summer months. The former mayor of Houston is quoted as saying that that fair city could not exist without air conditioning.[4] Most of the remaining energy used in the residential/commercial sector was employed for water heating, cooking, and refrigeration.

Industry

The heart of an industrial society is its industries. In 1970, three industries accounted for more than half the energy used in this sector: the metal, chemical, and oil-refining industries. Significantly, the food-processing industry was fourth in energy consumption, an issue to which we will return in Chapter 9. Nearly half (46.5%) the energy used to power U.S. industry was obtained from natural gas, and its use in the industrial sector was growing at the fastest rate.

Electric Utilities

The large utility companies that generate and sell electricity comprise the fastest growing sector of the economy in terms of energy use, reflecting the increasing dependence of industrialism on electricity. By 1990 the U.S. government expects the utilities to consume 39% of the total energy budget. Electricity is thermodynamically inefficient, because its utilization requires several energy conversions, with the attendant dissipation of energy at each step as mandated by the second law of thermodynamics. In an age of abundant energy, however, such inefficiency seemed trivial. The ease and convenience with which electricity can be transported and controlled has made it central to industrialism in the U.S. and elsewhere.

Fossil-Fuel Production Cycles: The Exponential Model

Not only are industrial nations more than 95% dependent on fossil fuels (Table 3.1); in addition, their use of fossil fuels is increasing at an exponential rate. This fact has been used by others—notably Meadows et al. in their well-known book *Limits to Growth*[5]—to project the future of industrialism. Partially in response to *Limits to Growth*, the exponential model has become more generally familiar in recent yers. In particular, it is widely appreciated that exponential growth is an exhilarating process[6] with at least three general defining features. First, the rate of exponential growth is usefully expressed by its doubling time, i.e., the amount of time it takes for a growing quantity to double in size. Second, this doubling time is the same at any stage of the growth process; thus, if it takes one year to double from 1 to 2, then at the same rate of exponential growth it also take one year to double from 67 million to 134 million. Third, exponential growth occurs when any quantity increases at a constant rate or annual percentage. The doubling time dcreases disproportionately as the growth rate in percent increases (Table 3.2). Thus if a bank account increases in size at 1% per year, it takes 69 years to double; but at 7% per year, the account doubles in a mere 10 years. Of course if inflation proceeds at the same rate, there is no net gain in purchasing power.

Since the use of fossil fuels has increased in recent decades at an exponential rate, one way to assess the future of industrialism is to compare these rates with the remaining reserves of each fuel. This technique for

Table 3.2. THE RELATION
BETWEEN ANNUAL PERCENT
INCREASE AND DOUBLING TIME
FOR AN EXPONENTIALLY
GROWING PROCESS.

Annual rate of increase (%)	Doubling time in years
0.5	138
1.0	69
1.5	46
2.0	35
2.5	28
3.0	23
3.5	20
4.0	17
4.5	15
5.0	14
5.5	13
6.0	12
6.5	11
7.0	10

analyzing the future is called extrapolation; let us apply it to each of the major fossil fuels.

Coal

For the eight centuries from 1060 to 1860 an estimated 7 billion metric tons of coal were extracted from the earth and burned. In the 90 years from 1860 to 1970, 132 billion tons were mined, nearly 19 times the cumulative production of the preceding 800 years.[7] We can calculate how long it will take to exhaust the world's reserves of coal if we know how large the remaining reserves are. These data are available, but they must be considered approximate. Estimates by different individuals or companies are based on different assumptions, and vary by a factor of at least four. According to liberal estimates, there are 7.6 trillion metric tons of coal left on earth, about 400 billion tons of which are located in the U.S. and recoverable. From the time the first ton of coal was mined nine centuries ago, the reader may calculate that 43 doublings are possible before world reserves are exhausted. By the end of 1970, 37 doublings had already occurred, leaving 6 doublings for the future. Assuming a very modest worldwide increase in the rate at which coal is mined and burned, say 3.5%, the doubling time shown in Table 3.2 is 20 years. Hence, at this modest growth rate six doublings will occur in 120 years. In other words, if exponential growth of coal use continues at this slower rate, world reserves will be exhausted in just over a century. If actual coal reserves are twice our estimated value, one more doubling will be possible, extending the use of coal by 20 years. This exercise illustrates that calculations based on exponential growth are insensitive to large errors in estimated reserves. But if historical growth rates of energy conversion of 5% per year are sustained indefinitely, then the doubling time is 14 years (Table 3.2), and world coal reserves will disappear within 84 years. This exercise demonstrates that calculations based on exponential growth are relatively sensitive to assumptions about growth rate, which is one of their main weaknesses.

Oil

Oil burns cleaner than coal, and although energetically more expensive to obtain, it is much easier to handle once it reaches the surface. Accordingly, and as illustrated in Table 3.1, oil has become the dominant fuel of industrial societies. In 1880, world production was barely significant, but from 1890 to 1970 world oil production grew exponentially at a rate of 7%,[8] corresponding to a doubling time of 10 years (Table 3.2).

According to experts, world reserves of oil are more difficult to estimate than coal, and hence somewhat more subject to error. We have seen, however, that exponential growth is relatively tolerant to large errors in estimating remaining reserves. According to one estimate, the world's initial produceable oil resources amounted to 2 trillion barrels.[9] The number of doublings required to exhaust this initial quantity is 41, of which 38 had already occurred by 1970. Thus, three doublings remained. At the historic annual growth rate of oil production of 7%, the corresponding doubling time is 10 years (Table 3.2). Therefore, according to this exponential model the world's initial produceable reserves of oil will be exhausted 30 years from 1970, or in the year 2000. Even if actual reserves are twice our estimate, only 10 additional years of oil production are gained.

In the U.S., oil became the dominant fuel near the end of World War II. Estimates of how much recoverable oil the U.S. has vary from a low of 150 billion barrels to a high of 400 billion barrels. Oil production in the U.S. increased exponentially from 1860 to 1970 at an annual growth rate of 7%, to a cumulative production of nearly 100 billion barrels by 1970. Under the assumption of continued exponential growth, it can be calculated that the U.S. has no more than 20 years of oil production remaining before its resources are totally exhausted.

Natural Gas

Natural gas is the most prized of fossil fuels. It is easy to obtain, flowing up from wells "automatically" under the pressure of oil deposits with which it is commonly associated. Once it reaches the surface, natural gas is exceptionally easy to handle and transport. Above all, natural gas burns clean, yielding little more than heat, carbon dioxide, and water.

Prior to W.W. II, facilities for handling natural gas were nonexistent, and most of it was simply burned on the well site. The value of natural gas was recognized, however. Beginning shortly after W.W. II, pipelines were built and reliable statistics recorded. Estimates of reserves are still subject to uncertainty, but for the U.S. they range from 900 to 1,300 trillion cubic feet, including Alaskan reserves. Let us assume an intermediate value of 1,100 trillion cubic feet. From W.W. II to 1960 the annual growth rate of natural gas use was 7%, corresponding to a doubling time of 10 years.[10] In 1960, 11 trillion cubic feet were produced. From these data, one can calculate that 3.2 doublings can occur from 1960 until existing U.S. reserves are exhausted, corresponding to 32 years. That is, the U.S. supply of natural gas will, by this calculation, be exhausted in 1992.

The Last Easter Egg

While the implications of exponential growth are widely appreciated, it is also increasingly appreciated that exponential growth does not provide a realistic model for resource depletion. Exponential growth can never be sustained indefinitely in a finite world, if only because resources become increasingly difficult to find as they are exhausted. Thus, although the positive feedback that underlies exponential growth is by nature self-reinforcing in its earliest stages (Chapter 2), it eventually becomes self-limiting, a process known to engineers as negative feedback. Perhaps the chief value of the exponential growth model lies in the very fact that it cannot furnish a realistic description of the future. It follows that the exponential growth of fossil-fuel use that has characterized the first three-quarters of the 20th century cannot be sustained, and that a change in the pattern of fossil-fuel use by industrial nations is inevitable. To understand the likely direction that this change will take clearly requires a more realistic view of the dynamics of resource depletion. Let us seek an intuitive understanding of depletion dynamics using the metaphor of an Easter egg hunt. Our "resource" in this case is the number of Easter eggs that remain to be found, but the metaphor applies equally well to fish in the sea, trees in a forest, and fossil fuels beneath the surface of the earth.

Suppose we are confronted with the task of finding a finite number of Easter eggs hidden in a meadow of finite dimensions. The likelihood of finding an egg in a given period of time depends principally on three variables: the number of eggs (or more precisely, their density in a finite area), how well the eggs are hidden, and how hard we look.

The Density of Eggs

Assuming that all eggs are equally well hidden and we search with constant effort, then it is self-evident that the higher the initial density of eggs, the more we shall find per unit time. As eggs are found, however, their density automatically declines, and the recovery per unit time also goes down. Under such idealized circumstances, it is easy to demonstrate that the number of eggs found per unit time declines exponentially (Figure 3.3a). The last Easter egg could no doubt be found, but it might well take more effort than justified by the value of the egg.

Quality of Concealment

Now let us add a complication to our hunt. Let us suppose that some eggs are hidden better than others, and are correspondingly more difficult to find. Under these circumstances, the most obvious eggs will clearly be most vulnerable to discovery, but as these eggs are found, two things happen. First, the density of eggs declines precipitously; and second, the difficulty of the hunt automatically increases because the remaining eggs are better hidden. The result is to exaggerate the initial recovery of eggs, with a corresponding decline in recovery later on (Figure 3.3b).

Search Effectiveness

Now suppose we vary the effectiveness with which we search for the eggs. Such variation is inevitable, if only because our skills increase with practice. At first we find few eggs, but as we learn and our technique improves accordingly, recovery increases until we encounter the limits set by the first two considerations, namely, declining density and increasing difficulty of finding eggs that are better concealed. By adding this third condition, our egg-recovery curve will increase steadily at first, but then reach a peak and decline owing to the first two conditions. The result of these interacting forces is a bell-shaped recovery curve (Figure 3.3c). The increasing difficulty of finding eggs on the declining phase of the curve can at first be offset by simply searching more diligently, or perhaps enlisting the aid of friends. The inevitable result, however, is that the eggs remaining to be discovered decline still faster. Eventually, the search is narrowed to the last Easter egg.

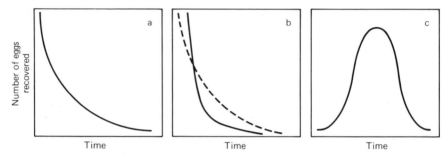

3.3   Theoretical recovery curves of eggs in an Easter egg hunt. Curve *a* illustrates the exponential decline over time expected if the quality of concealment and the search effectiveness are constant. Curve *b* shows the effect on recovery of hiding some eggs better than others (solid curve), together with the exponential decline in *a* for comparison (dashed curve). Curve *c* shows the additional effect of variation in search effectiveness (see text).

Fossil-Fuel Production Cycles: The Bell-Shaped Model

The general rules of resource recovery illustrated by the metaphorical Easter egg hunt apply fully to the production of fossil fuels. Thus it is unrealistic to expect fossil-fuel production to grow exponentially until it suddenly drops to zero, as implied by the exponential model. Instead, fossil-fuel recovery may be expected to build to a peak and then decline, in a bell-shaped production curve resembling the recovery curve of Easter eggs (Figure 3.3c).

The dynamics of oil recovery furnish an intuitive basis for understanding the bell-shaped production cycle as applied to fossil fuels. During the ascending phase of oil production, positive feedback reigns.

Knowledge of how to find and produce oil increases, and the energy stored in the extracted oil enables the construction of machinery and the accumulation of capital that allows yet more oil to be produced. During this ascending phase, the production cycle is governed largely by the historic pattern of the feedback of energy to make still more energy available (Chapter 2). As production approaches the peak, however, new forces predominate. Owing to the physics of oil flow to the well through the porous rock in which the viscous substance is generally trapped, the flow rate of any given well or field declines exponentially over a predictable period of several years. As a general rule no more than 10% of an oilfield's total reserves can be produced in one year without irreversibly reducing the quantity of oil that can ultimately be recovered.[11] Therefore, increasing overall production of oil requires continual new discoveries; but as the resource becomes exhausted, the rate of new discoveries declines and the production rate follows suit. Recovery slows, and the production curve peaks. If no new oilfields are developed, the overall production curve reflects the summed declining exponential of all the curves for individual wells and fields, accounting for the downward leg of the bell-shaped curve.

Secondary and tertiary recovery techniques can "wring" the earth of its remaining reserves, prolonging the decline somewhat. Secondary production involves pumping water (generally brine) into the oilfield to raise the pressure and force oil out the wells, while tertiary recovery entails the injection of steam or chemicals which make the oil less viscous and therefore easier to pump.[12] These techniques are energetically expensive, however, in comparison with primary production. Their very use is a sure sign that the resource is becoming exhausted. Eventually more energy is required for recovery than is returned in oil produced, and recovery accordingly ceases.

A consequence of this depletion dynamic that is not yet widely appreciated, and to which we shall return, is that long before a resource is exhausted its rate of recovery peaks and begins to decline. It is inaccurate to suppose we will "run out" of energy. Long before the last barrel of oil is extracted from the earth it will become uneconomic to remove that barrel. It will take more energy to extract the oil than is contained within the oil itself, in which case extracting the oil will incur a net loss of energy available to industrial society. The most relevant portion of the resource recovery cycle is the time of peak production. Before that time, everything in industrial societies expands: the economy, the size of automobiles, and population. After the time of peak production, everything contracts. Thus, the key to the future of industrialism lies in the timing of the production peak of energy and natural resources. Fortunately, the mathematics of the bell-shaped resource production cycle of fossil fuels have been analyzed by geologists (mainly M. King Hubbert),[13] allowing the time of peak production to be determined with some precision. Let us examine the implications for each of the major fossil fuels on which industrial societies depend.

Coal

We saw earlier that the U.S. possesses 200–400 billion tons of coal. With present patterns of coal use in the U.S., the production of coal may be expected to peak in 150 years (Figure 3.4). Because the the relative disadvantages of coal, its use has declined steadily in the U.S. since W.W. II, and it now accounts for about 20% of the gross fuel input to U.S. industrial society. If we rely increasingly on coal as a source of fuel, however, which seems likely, then peak production will occur much sooner, perhaps in the first half of the next century. As we shall see in future pages, however, social, economic, and environmental pressures may constrain the use of coal well below the maximum implied by remaining reserves.

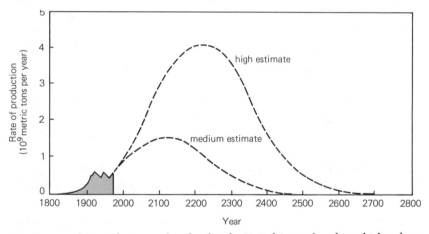

3.4 Two complete production cycles of coal in the United States, based on a high and medium estimate of the initial amount of recoverable coal. The medium estimate (390 × 10⁹ metric tons) is probably the more realistic. Shaded area shows actual cumulative production to date. (after M. K. Hubbert, *U.S. Energy Resources, A Review as of 1972*, Document #93–40 [92–75] [Washington, D.C.: U.S. Government Printing Office, 1974], Figure 22)

Oil

Oil is by far the most important fuel for industrial societies at present (Table 3.1). As we have seen, estimates of how much oil the U.S. has both in production and yet to be discovered vary widely, from 150 to 400 billion barrels. Calculations based on the conservative figure of 170 billion barrels yield the production cycle or recovery curve shown in Figure 3.5. This estimate suggests that oil production in the U.S. will peak sometime before 1980, and decline to half the present value by 2010. The Alaskan reserves show up as a small hump on the declining leg of the recovery curve. The exponential decline of Alaskan production will sum with the decline in production in the contiguous United States around 1990, little more than a decade away.

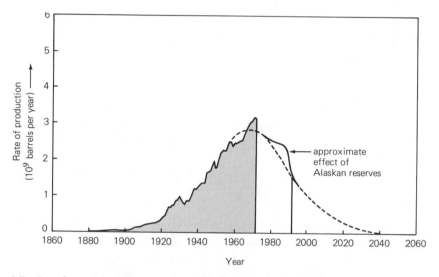

3.5 Complete production cycle of crude oil in the conterminous United States, based on an estimate of 170×10^9 barrels. The shaded area shows actual production to date. (after M. K. Hubbert, *U.S. Energy Resources, A Review as of 1972*, Document #93–40 [92–75] [Washington, D.C.: U.S. Government Printing Office, 1974], Figure 51)

Natural Gas

As we have seen, the U.S. is especially dependent on natural gas to fuel its industrial economy (Table 3.1). Thus the future of natural gas is especially relevant to U.S. industrialism. Computed production cycles for natural gas, based on estimates of known plus undiscovered reserves, are shown in Figure 3.6. Production in the U.S. has already peaked and is now declining. As in the case of oil, the Alaskan production may be expected to appear as a small bump on the declining phase of the curve. By 1990, the exponential decline of Alaskan production will sum with the exponential decline in the rest of the U.S.

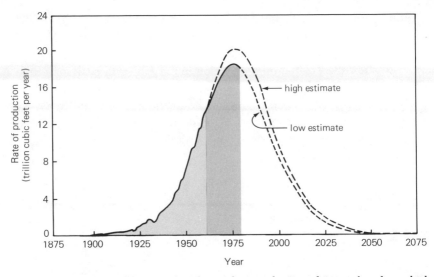

3.6 Two complete production cycles of natural gas in the United States, based on a high and low estimate of the initial recoverable gas. Shaded area shows actual production to date. (after M. K. Hubbert, *U.S. Energy Resources, A Review as of 1972*, Document #93–40 [92–75] [Washington, D.C.: U.S. Government Printing Office, 1974], Figure 53)

Global Projections

So far we have focused on the energy future of the U.S. In fact, this nation has been the world's most prolific producer of crude oil, but as of 1975 it possessed only 6% of the world's remaining proven reserves.[14] Moreover, and as we shall discuss shortly, the U.S. depends increasingly on foreign sources of oil to satisfy its voracious energy appetite. A full appreciation of our energy future thus requires that we adopt a global perspective.

Among the nations of the world, the Soviet Union is the world's second-largest consumer of commercial energy, and in 1975 became the world's leading producer of crude petroleum.[15] Because of the Soviet Union's soaring demand, however, its most productive oilfields are beginning to lag.[16] As a result the nations of Eastern Europe, long dependent on Soviet oil, are having to turn elsewhere for their supply, notably to Iran,[17] and gasoline prices within the U.S.S.R more than doubled in 1978.[18] China, the third-largest consumer of commercial energy,[19] may have vast undiscovered offshore deposits of oil in the China Sea[20]; but its single most productive oilfield, from which most of its petroleum is derived, is on the verge of rapid decline.[21]

Of the remaining 658 billion barrels of proven worldwide reserves,[22] the Organization of Petroleum Exporting Countries (OPEC) commands fully 68%, or 450 billion barrels. Saudi Arabia alone has 23% of the total remaining world oil, or 152 billion barrels. These figures seem huge until they are compared with the cumulative world production as of 1975 of 341 billion barrels.[23] To put it differently, cumulative world oil production as of 1975 equalled more than half the proven remaining reserves in the non-Communist world, and world production (=consumption) is still soaring. Although OPEC is the most important future source of U.S. oil, its production is projected to peak sometime between 1995 and 2015 unless production is held near current levels by deliberate government regulation.[24] A recent, two-year international study by scholars and business representatives, encompassing both "optimistic" and "pessimistic" assumptions about growth rates and additions to reserves, concluded that the world demand for oil will outstrip the supply sometime between 1995 and 2015.[25] If OPEC deliberately constrains production to husband its remaining reserves, which is possible,[26] then demand will exceed the supply even sooner. In view of the declining domestic and global availability of both oil and natural gas, the final decade of the 20th century may mark a monumental watershed for the U.S. and for industrial civilization in general.

Hubbert
Versus the Giants

M. King Hubbert, a renegade geologist, was principally responsible for the foregoing analysis of U.S. fossil-fuel production cycles. In the 1950s, when U.S. industrialism was flexing its muscles in preparation for the phenomenal growth of the 1960s, Hubbert stood almost alone in foreseeing the decline of fossil-fuel availability. It is perhaps not surprising that his views were greeted first with incredulity, then polite indifference, and finally with scorn and ridicule. Oil companies were especially indignant; they have an obvious emotional and economic stake in maintaining the production of fossil fuels, and it is perhaps understandable that their judgments were colored by this hope. The oil companies responded to Hubbert's analysis first by increasing their estimates of undiscovered oil reserves. Hubbert plugged their new and higher estimates into his equations and showed that the time of peak production was delayed by only a few years to decades.

Time passed, and the curve of actual oil production faithfully paralleled Hubbert's predictions (Figure 3.5). Moreover, the discovery of new reserves began to level off. Hubbert reasoned that following the discovery of a reserve, its development into a productive well would require a few years. Thus the production peak should lag the discovery peak by a specific amount of time, calculated by Hubbert as 16 years for natural gas and 11 years for oil.[27] The rate of natural gas discovery in the U.S. peaked sometime between 1950 and 1960 and has declined since. Proved discoveries of oil in the conterminous U.S. peaked in 1958 and have declined since. In accord with Hubbert's analysis, the actual production curves of both oil and natural gas in the U.S. have peaked and are now declining.

Faced with these facts, the oil companies had little choice but to try a new approach. They accepted Hubbert's data, but argued for a fundamentally different interpretation. Declining production, they asserted, was related not to the depletion of the resource, as Hubbert suggested, but rather to the lack of economic incentive for exploration. If the government would only allow the oil companies to sell their product at a higher price, they would have the funds needed to produce more oil and more natural gas. Hubbert responded to this position by examining the success of exploratory drilling in the U.S. on a year-by-year basis over the past several decades. If the oil companies were correct, then the discoveries per foot of exploratory drilling should have remained relatively constant at some high level. If, on the other hand, the decline in production is truly attributable to the depletion of the resource, then oil should be harder and harder to find. That is, cumulative discoveries should level off as cumulative exploratory drilling increased. When the data were graphed, Hubbert was again vindicated (Figure 3.7); since 1945, exploratory wells have become less and less productive. Four decades ago, each foot of well hole drilled yielded 275 barrels of oil; today, each foot yields but 25 barrels.[28] More "dry wells" are now being drilled and the depth of productive wells is increasing steadily. The simplest interpretation of these facts is that oil productivity is declining because the supply remaining in the ground is diminished. As oil becomes depleted, it indeed is more costly to extract in energetic and, therefore, economic terms; thus the oil companies may indeed need to charge more for their product in order to maintain their profits. Hubbert is vindicated, however, in that his predictions of the 1950s have been validated by the events of subsequent decades.

No one, including Hubbert, denies that estimates of remaining fossil fuels are uncertain and that the U.S. has substantial oil yet to be discovered, most in Alaska and offshore beneath the outer continental shelf. Neither, however, does anyone deny the essential correctness of the

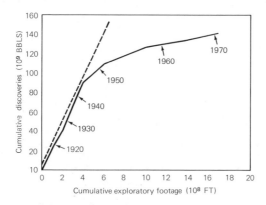

3.7 Cumulative oil production in the United States plotted against the depth of exploratory drilling. If reserves are not being depleted, the relation between these two variables would form a straight line as indicated by the dashed curve. The solid curve shows actual discoveries, which have been declining steadily since about 1945 and are now approaching zero (as the curve approaches a horizontal line). (after M. K. Hubbert, *U.S. Energy Resources, A Review as of 1972*, Document #93–40 [92–75] [Washington, D.C.: U.S. Government Printing Office, 1974], Figure 48)

Hubbert analysis. Mobil Oil Company now predicts the "exhaustion" of U.S. oil by the year 2000; the National Academy of Sciences, 2003; the National Petroleum Council, 2007; the U.S. Geological Survey, 2012 (low estimate) to 2028 (high).[29] All of these figures are based on a 2.5% annual growth rate of energy use. As we have seen, the actual worldwide rate of increase from 1945 to the present was 4.38%,[30] and for oil in the U.S. the growth rate was near 7%.[31] Unless we curb our appetite for oil, therefore, we shall exhaust domestic supplies well before 2000.

King Hubbert's Hump

What then is the prognosis for the future of industrialism? From the admittedly limited vantage point of the present, it would appear that within a few decades to a half century the world's major stores of fossil fuels will have become depleted beyond usefulness. In the U.S., oil and natural gas production will decline steadily through the remainder of the 20th century. Coal production may be expected to increase steadily, reaching a peak sometime in the next century. When fossil fuel use is seen in broad perspective, it may appear as a small hump on the otherwise smooth scale of history (Figure 3.8).

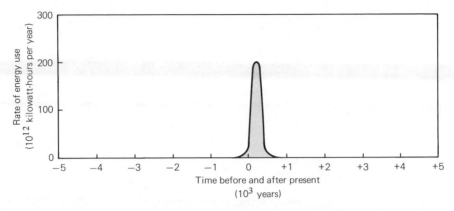

3.8 Probable rate of fossil fuel use as seen on the broad time scale of human history, from 5,000 years ago to 5,000 years into the future. (after M. K. Hubbert, *U.S. Energy Resources, A Review as of 1972*, Document #93–40 [92–75] [Washington, D.C.: U.S. Government Printing Office, 1974], Figure 69)

Short-Term Solutions

There is no longer any disagreement that the U.S. is rapidly exhausting its supplies of fossil fuels. The two fossil fuels that are in shortest supply, oil and natural gas, provided an average of 68.8% of the energy needs of the major industrial powers in 1970 (Table 3.1). In the U.S., 77.5% of the energy budget was supplied by these two fossil fuels, with 34.8% coming from the scarcest fuel, natural gas. There is virtually universal agreement that within 25–50 years the fuels that presently drive from two-thirds to three-quarters of the world's industrial activity will no longer be available in significant amounts. Clearly, a major change is imminent.

The United States, as we have seen, is already experiencing a decline in production of oil and natural gas, and yet projected industrial activity calls for a doubling of energy conversion over the next two decades. Where will the energy come from? One source is foreign oil and natural gas. "Project Independence" notwithstanding, the United States will, in the short term, become steadily more dependent on foreign oil. Already the most advanced industrial nations—the United Kingdom, West Germany, and Japan—are nearly totally dependent on imported energy. In 1975, 40% of the oil used by the U.S. was imported, mostly from OPEC. By 2000, this dependence is projected by the U.S. government to reach 72% (Fig. 3.9). Similarly, in 1975 the U.S. imported 8% of its natural gas;

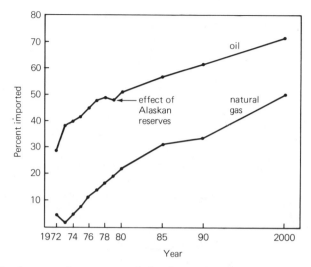

3.9 Actual and projected importation of oil and natural gas into the United States. (based on a U.S. Government study reported in P. L. Auer, "An Integrated National Energy Research and Development Program," *Science* 184 [1975]: 295–301)

by 2000 the dependence is projected by the U.S. government to increase to 50%.

For the short term, then, industrial nations may be able to satisfy their appetite for energy from foreign sources. We have seen, however, that the depletion of fossil fuels is not confined to the U.S.; rather, it is a worldwide phenomenon. Importation of fossil fuels is clearly a stop-gap measure that cannot be effective for longer than a decade or two. Never mind the monumental political implications of the dependence of the United States on Mideast oil, which has already transformed U.S. Mideast policy; never mind the profound economic implications of the cash drain associated with buying foreign oil, which is already shaking the world economy to its foundation; and never mind the increasingly strident demands for fossil fuels by nonindustrial nations, which will compete increasingly for remaining supplies. The fact is that the U.S. and other industrial nations cannot depend long on foreign oil simply because that oil will not be available for long. The oil-importing nations have an estimated 10–20 years to accomplish the transition from oil to other energy sources.[32] The continuation of industrialism as we have known it therefore depends entirely upon the possibility of employing alternate sources of energy. It is to these alternatives that we now turn.

NOTES

1. M. K. Hubbert, *U.S. Energy Resources, A Review as of 1972*, Document #93–40 (92–75) (Washington, D.C.: U.S. Government Printing Office, 1974), p. 45.
2. K. Teltsh, "Energy Output Tripled in 25 Years, U.N. Says," *New York Times*, 30 July 1976, p. D1.
3. G. A. Lincoln, "Energy Conservation," in *Energy: Use, Conservation and Supply*, edited by P. H. Abelson (Washington, D.C.: American Association for the Advancement of Science, 1974), pp. 19–26.
4. J. P. Sterba, "Houston, as Energy Capital, Sets Pace in Sunbelt Boom," *New York Times*, 9 February 1976, p. 1.
5. D. H. Meadows, D. L. Meadows, J. Randers, and W. W. Behrens III, *The Limits to Growth* (New York: Signet, 1972).
6. L. R. Brown, *The Twenty Ninth Day* (New York: W. W. Norton, 1978).
7. Hubbert, *U.S. Energy Resources*, p. 27.
8. Ibid.
9. Ibid.
10. Ibid.
11. A. R. Flower, "World Oil Production," *Scientific American* 238 (1978): 42–49.
12. Ibid.
13. Hubbert, *U.S. Energy Resources*.
14. Flower, "World Oil Production"; National Academy of Sciences (1975). "Mineral Sources and the Environment" (Washington, D.C.: National Academy of Sciences, 1975); U.S. Geological Survey, "Geological Estimates of Undiscovered Recoverable Oil and Gas Resources in the United States," Geological Survey Circular 725 (Washington, D.C.: U.S. Geological Survey, 1975); Federal Energy Administration, "National Gas Emergency Standby Act of 1975," Report #DES 75–5 (Washington, D.C.: Federal Energy Administration, 1975); M. K. Hubbert, "The Energy Resources of the Earth, *Scientific American* 224 (1971): 60–70; A. B. Lovins, "World Energy Strategies," *Bulletin of the Atomic Scientists*, May 1974, pp. 14–32; id., "The Case for Long-term Planning," *Bulletin of the Atomic Scientists*, June 1974, pp. 38–50; N. Grove, "Oil: The Dwindling Treasure," *National Geographic* 145 (1974): 792–825.
15. Teltsh, "Energy Output Tripled."
16. S. Rattner, "Signs Appear That Soviet Supply of Oil is Tightening," *New York Times*, 21 November 1977, p. 53; D. K. Shipler, "Soviet Is Facing Harsh Oil Task," *New York Times*, 24 January 1978, p. 1.
17. D. A. Andelman, "Iran Emerging as Major Alternative in Eastern Europe's Search for Oil," *New York Times*, 21 November 1977, p. 53.
18. C. R. Whitney, "Soviet Sharply Increases Price of Gas and Coffee," *New York Times*, 2 March 1978, p. A3.
19. Teltsh, "Energy Output Tripled."
20. S. S. Harrison, "Time Bomb in East Asia," *Foreign Policy* 20 (1975): 3–27; C-

h. Park and J. A. Cohen, "The Politics of the Oil Weapon," *Foreign Policy* 20 (1975): 28–49.

21. R. H. Munro, "Production of China's Oilfield at Taching May Have Peaked," *New York Times*, 27 December 1976, p. D2.

22. Flower, "World Oil Production."

23. Ibid.

24. Ibid.

25. Ibid.; E. Cowan, "World Oil Shortage Is Called Inevitable," *New York Times*, 17 May 1977, p. 1; C. L. Wilson, *Energy: Global Prospects, 1985–2000* (New York: McGraw Hill, 1977).

26. S. Rattner, "Saudi Increase in Output Seen by Schlesinger, But Rise May Not Match World Oil Demand," *New York Times*, 25 January 1978, p. 1.

27. Hubbert, *U.S. Energy Resources.*

28. E. T. Hayes, "Energy Implications of Materials Processing," *Science* 191 (1976): 661–65.

29. D. Shapley, "Senate Study Predicts U.S. Oil 'Exhaustion,'" *Science* 187 (1975): 1064.

30. Hubbert, *U.S. Energy Resources*, p. 45.

31. Ibid., p. 27.

32. Flower, "World Oil Production"; Andelman, "Iran Emerging . . ."; Harrison, "Time Bomb in East Asia"; Park and Cohen, "Politics of the Oil Weapon.'

Chapter 4
ALTERNATIVES
TO CONVENTIONAL
FOSSIL FUELS

There is broad agreement that the traditional fossil fuels, notably oil and natural gas, will become steadily more scarce and finally exhausted in the lifetime of most people now alive on earth. In particular, the critical function of transportation is almost fully dependent on the liquid fuels whose depletion is most imminent. Without a modern transportation system the products of factories could not be brought to market, people could not reach the workplace, and foodstuffs could not be distributed from rural production sites to urban consumers. In short, industrialism as we know it would cease without a relatively inexpensive and rapid system of transportation, which in turn depends on nonsubstitutable liquid fuels. The governments of industrial nations realize increasingly that their current economies and lifestyles depend on the development of alternative fuels.* What are these alternatives, and how likely is it that they can be developed soon enough to compensate for the dwindling supplies of oil and natural gas?

Synthetic Fuels

One alternative to conventional fossil fuels is the "synthetic" fuels that are made from coal. If we consider only the issue of available supplies, there

---

* In the U.S., for example, the Energy Program announced by the Carter administration in April 1977 asserts that "America's hope for long term economic growth beyond the year 2000 rests on renewable and virtually inexhaustible sources of energy, such as solar and geothermal energy."[1]

is enough coal to satisfy the energy appetites of industrial nations well into the next century. Coal cannot be used directly in place of oil and natural gas, but by converting it to synthetic liquid fuels coal can in principle substitute partly for the dwindling oil on which modern transportation almost entirely depends. Likewise, by converting coal to synthetic gas it can be used in place of the natural gas that now fuels half of all U.S. industry.

The Liquification of Coal

Techniques for converting coal to gasoline have been known at least since World War II, when the Germans were forced to develop such methods to fuel their military machine. To liquify coal, it is dissolved in an organic solvent and then filtered to remove impurities that would otherwise clog engines and foul the air. The resulting liquid can be burned directly in place of coal with a few modifications to existing boilers. Alternatively, the dissolved coal can be processed further to produce a synthetic crude oil similar to natural oil. This synthetic oil can then be refined into gasoline and other liquid fuels.[2]

Coal liquification is simple in principle, but in practice the process is awkward and to date uneconomic. As yet,

> there is no commercial industry in place today for making liquids directly from coal, nor, for that matter, is there technology available for producing liquids from coal at an economically competitive price.[3]

Moreover, the development and commercialization of a liquified coal technology faces a number of obstacles that are unprecedented in the history of the energy industry. In particular, the research and development costs are unusually high, and the timing, conditions, and economics of future commercial applications are unclear.[4] These obstacles represent added challenges to the development and commercialization process, and suggest to some observers the desirability of large government incentives to private industry.[5]

The energy industry maintains a moderate research program for synthetic fuels, and improvements in existing technology may be forthcoming. Contracts have been let and construction begun on the first large-scale experimental coal-liquification plant in Kentucky, scheduled originally for completion in 1978.[6] The prospects are sufficiently unpromising, however, that the Energy Research and Development Administration of the U.S. government has "dropped outright its goal of regular commercial production of oil from coal by 1985,"[7] largely because synthetic fuels are so difficult to produce that they cannot compete economically with other fuels.[8] The U.S. Government Accounting Office has recommended instead an increased dependence on foreign oil.[9] Political winds are always shifting and coal liquification may yet receive priority research funding from the U.S. government. In the meantime it is unrealistic to expect that the supply gap left by declining petroleum can be filled by coal liquification.

Coal Gasification

Coal can be converted to gas by pumping hot air or a mixture of oxygen and steam through it. The process requires several steps and yields various grades of combustible gas, including some pipeline-quality natural gas. Several alternative coal gasification technologies are now available,[10] and the first pilot coal-gasification plant in the U.S. commenced operation in 1977 in New York State. The gas produced in this pilot plant contains but one-sixth the energy value of natural gas, but it is relatively cheap to make and it may prove economic to market.[11]

At the present time, coal gasification seems the most feasible alternative to conventional fossil fuels. Technical problems remain, but they seem comparatively minor.[12] One source estimates that 15% of the projected gas requirements of the U.S. could be met from coal by 1985.[13] This estimate is probably unrealistically high, but it is safe to predict that an increasing proportion of the gas burned by industrial societies will be produced synthetically from coal. The inevitable result, however, will be to accelerate the depletion of existing coal reserves. Moreover, and as we shall explore in future chapters, there are major technical, environmental, and institutional limitations to the expanded use of coal. Finally, coal gasification cannot produce the liquid fuels on which modern industrialism depends. Coal gasification may be expected at most to serve as a partial bridge to a different energy future.

Oil Shale and Tar Sands

As we have seen, most of the world's oil is trapped in underground formations of porous rock. Some oil, however, is found in association with shale rock and in "tar sands." The major shale rock deposits are located in the western U.S., notably in Colorado, Utah, and Wyoming. Tar sands are located almost exclusively in Alberta, Canada.

According to frontier legend, shale rock was discovered accidentally by an early pioneer, who had used such rock to construct his fireplace. As the pioneer quickly learned, the rock is flammable. When it burns, it releases a sticky, tarlike substance called kerogen, which can then be refined to crude oil and its products. The process is beset with severe technical and environmental limitations, however, not the least of which is where to put the countless tons of "waste" rock after the kerogen has been extracted. The most viable method of disposal from an economic standpoint is to dump the rock into empty canyons, filling them to the brim and threatening to leach mineral impurities into water supplies. Needless to say, this proposal is not received lightly by environmentalists nor agriculturalists.

The total oil reserves contained in shale are estimated at 50–100 billion barrels, a significant quantity in comparison with 200–400 billion barrels contained in conventional underground oil deposits in the U.S. The limitations on recovering shale oil are so severe, however, that the Department of the Interior, the Atomic Energy Commission, and industrial representatives estimate that at most 1–2 million barrels a day are recoverable.[14] This figure pales beside the daily U.S. consumption of 18 million barrels.[15] These and other limitations on the economic use of shale rock have induced the U.S. to suspend its only prototype shale rock development program in Colorado.[16]

Oil contained in the tar sands of northern Alberta can be recovered economically, although the process is slow and incredibly demanding of workers and machines. Each winter the ground freezes, forcing recovery operations to a halt. Production resumes in the spring after the bulldozers are rescued from the thawed mire into which they frequently sink. In addition to the recovery problem, the technology for extracting oil from the viscous sands is not well developed. Such problems may be academic, however, since Canada has announced its intention to halt all energy exportation to the U.S. by 1982.[17] Even under optimal technical and political conditions, both shale rock and tar sands can supplement, but not replace, the dwindling oil and gas stocks on which industrial civilization is based. In the judgment of at least one industry representative,

> it is unlikely that synthetic crudes will ever be the dominant feedstock for petrochemicals. Rather, we see such materials as . . . extending the duration of . . . the era of oil and gas.[18]

We must therefore turn elsewhere in our search for prospective fuels for the future.

Nuclear Power

In its steady march toward the use of more concentrated fuels, humanity has at last reached the pinnacle. Science knows of no source of energy that is more concentrated than that which holds atomic nuclei together, and modern technology has begun to tap this energy for human use. Extracting energy from atomic nuclei involves converting a small amount of matter to an enormous quantity of energy, in accord with Einstein's famous equation, $E = MC^2$. This equation states that energy and mass are interchangeable.* If one gram of matter, about the size of a small marble, were completely converted to energy, it would yield the energy equivalent of 20,000 tons of TNT.[19]

As mentioned earlier (Chapter 1), nuclear energy can be extracted in the form of heat by one of two different processes: fission or fusion.

Fission

As we have seen, fission involves splitting the nucleus of one element (e.g., uranium) into the nuclei of two smaller elements, which together have less mass than the original uranium nucleus. The difference in mass is converted to energy in accord with Einstein's equation. Very few substances can undergo spontaneous fission. By far the most naturally abundant fissionable substance is uranium-235 (^{235}U), which is found in ore deposits in the earth's crust. When ^{235}U is extracted from raw uranium ore, purified, and then brought together in a large enough mass, the uranium nuclei undergo spontaneous fission in a positive-feedback or chain reaction, with the attendant release of energy (Figure 4.1).

The fission reaction can be controlled in power plants and used to generate heat from which electricity can be produced. The technology is already developed, as attested by the more than 200 fission reactors currently ordered, under construction, or actually operating in the U.S.[20] The proponents of ^{235}U fission realize, however, that this source of power is a temporary, stop-gap measure. Even if we ignore for the moment the environmental, political, and legal constraints on nuclear power (some of which we will discuss later), ^{235}U fission suffers a telling limitation: the scarcity of the fuel itself. If all the energy demands of industrial nations could be met beginning today with ^{235}U, the world's known stores of this substance would be exhausted well before 2000. ^{235}U is nearly as scarce as oil and natural gas.[21] Existing fission technology, like synthetic fuels, can

---

*E = energy, M = mass, and C = the velocity of light. The energy realized from a small mass is enormous because the mass is multiplied times the square of the velocity of light.

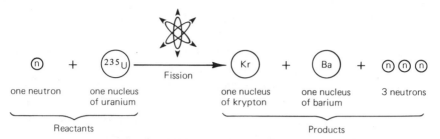

4.1 Nuclear fission of uranium, as occurs within a fission reactor. A nucleus of uranium (^{235}U) absorbs one neutron (N), splitting into krypton (Kr) and barium (Ba) plus three neutrons. The released neutrons then interact with another uranium nucleus to sustain the chain reaction. The mass of the reactants is greater than the mass of the products; the difference in mass is not actually lost, but is rather converted to energy and realized as heat. (modified from G. M. Crawley, *Energy* [New York: Macmillan, 1975], p. 111)

do little more than give industrial society a little additional time to make some more permanent energy arrangement for the future.

The Fast Breeder Reactor

Natural deposits of uranium contain only 0.7% of ^{235}U. The remaining uranium takes the form of uranium-238. Although ^{238}U is a close chemical cousin of ^{235}U, there is a critical distinction: whereas ^{235}U is capable of spontaneous fission, ^{238}U is inert. When it is brought together in a large mass, nothing happens. ^{238}U is, however, 100 times more abundant than ^{235}U. Moreover, when ^{238}U is exposed to a radioactive source it readily absorbs a neutron and is thereby converted to a new element, plutonium-239. ^{239}Pu is even more fissionable than ^{235}U, and can be used either to manufacture nuclear weapons or as fuel in a fission reactor.

In ordinary nuclear reactors, small quantities of ^{238}U are converted to ^{239}Pu. The fast-breeder reactor is designed to speed up this conversion in order to "breed" its own fuel. The fuel rods of the fast-breeder are blanketed with the inert ^{238}U, which is "automatically" converted to ^{239}Pu as fission takes place in the reactor. The plutonium is then removed and purified by a secret process. In reality, the fast-breeder reactor does not "breed" its nuclear fuel; it simply converts the abundant but inert ^{238}U into fissionable ^{239}Pu.

In principle, at least, the fast-breeder reactor is a straightforward way to extract energy from the relatively abundant and otherwise useless ^{238}U. In the process, the effective lifetime of existing uranium deposits could be extended 50- to 80-fold. In practice, however, the history of the fast-breeder reactor tells a much different story. Such power plants would cost so much to build (about $1.5 billion),[22] operate, and protect from sabotage that the electricity they generate might have to be sold at exorbitant prices. Simply to light a typical home, for example, could cost hundreds of dollars per year. A prototype fast-breeder reactor that the U.S. planned to construct in Tennessee was projected in 1973 to cost $700 million, but by 1975 cost overruns had escalated the actual cost to about $3 billion.[23] Moreover, cost estimates for the future have never adequately accounted for the processing of radioactive waste materials, nor included the environmental and social costs of the fast-breeder reactor.[24]

The commercial feasibility of the fast-breeder is the first of several major obstacles to its widespread use. In addition, the technology and safety of the fast-breeder are unproven. The first and only breeder reactor in the U.S., the Enrico Fermi plant in Detroit, suffered a disabling accident in 1966 in which the fission reaction went out of control and came within a heartbeat of causing a devastating nuclear explosion. Had the explosion occurred, it would have innundated Detroit with a deadly cloud of radioactivity, perhaps killed thousands of people, forced the evacuation of the city, rendered the region "permanently" uninhabitable (the half-life of plutonium-239 is 24,000 years), and effectively terminated the largest single U.S. industry—auto manufacture. The meltdown burned itself out without exploding, leaving in its wake a sobered scientific community and a twisted mass of intensely radioactive molten steel that is still being painstakingly dismantled.[25]

Such serious accidents have not yet occurred in Europe where research on the fast-breeder is more advanced.[26] Even with greater European experience, however, technical and safety problems remain unresolved,[27] and the commercial feasibility of the European breeder is uncertain. In particular, recurrent plumbing leaks have crippled the British and Soviet breeders for several years, and the French Phénix, a 250-megawatt prototype that has been considered a symbol of breeder reactor success,[28] was shut down in October 1976 for a several-month period of repairs for similar plumbing problems.[29] The experience with breeder reactors to date illustrates how minor technical problems may cause major interruptions in operation, reducing the "capacity factor" of the reactor and further impairing its commercial prospects.

The fast-breeder reactor suffers a more serious drawback, however, than technical viability or commercial feasibility. Each such reactor would require tons of plutonium fuel during its working lifetime—fuel that would have to be moved to and from distant reprocessing facilities over public transportation routes. Protection against a determined sabotage effort is impossible to guarantee, just as protection from million-dollar bank robberies cannot be assured. Plutonium can be wrapped in metal foil and carried in a briefcase, and only a few kilograms are needed to fabricate a Hiroshima-strength atomic weapon. Given the availability of plutonium, construction of such weapons is not difficult. Two years ago, a Princeton University undergraduate wrote a term paper showing how to make a nuclear weapon from plutonium and $2,000 of additional materials.* His only source of information was physics textbooks and public documents.[30] More recently, a 22-year-old former student with only one year of college physics designed a series of nuclear devices that government bomb makers termed "highly credible."[31]

Apparently the only serious obstacle between terrorist organizations and nuclear weapons is acquisition of plutonium. In fact, enough plutonium and uranium to produce scores of nuclear weapons has already disappeared from U.S. government installations. In 1976, for example, the federal government issued a classified report acknowledging its inability to account for 6,000 pounds of weapons-grade nuclear materials. In addition, the U.S. has lost track of "sizeable quantities" of weapons-grade atomic materials leased to foreign nations in the 1950s and 1960s.[32] The largest single known instance of nuclear theft occurred a decade ago, when an entire cargo ship laden with enough uranium ore to produce 133 plutonium warheads disappeared from the high seas. A few weeks later the ship reappeared—empty, with a new crew, and under a new flag. Nuclear pirating is not new. If plutonium is available in the massive quantities required by the fast-breeder reactor, the immense destructive power of the atom will be available to whomever is daring enough to seize it.

The foregoing drawbacks of the breeder reactor pale before its most frightening disadvantage: that the plutonium fuel that would be burned and generated by the fast-breeder is among the most toxic substances known. A few millionths of a gram of plutonium can cause lung cancer, and a chunk of plutonium the size of an orange would, if disseminated over the surface of the earth, exterminate most forms of life. Plutonium has already been detected in the environment near the Rocky Flats Nuclear Weapons Plant outside Denver, Colorado, and tons of radioactive wastes have leaked from storage facilities and found their way into the environment.[33] A 1976 report by the federal Nuclear Regulatory Commission acknowledged that existing designs for breeder reactors did not fully eliminate the possibility of "energetic core disruption" (that is, small

* Soon after his term paper made headlines, he allegedly received a call from a Pakistani official interested in buying his design.

nuclear explosions within the reactor), and such disruptions could inject vaporized plutonium into the atmosphere.[34] The report noted that the release of as little as 0.01% of the total plutonium contained in a typical breeder reactor (2,000 pounds) would be "cause for alarm."[35] Moreover, highway and railway accidents involving toxic industrial substances and even radioactive materials are not uncommon;[36] similar accidents in the transport of plutonium could conceivably threaten the entire human population. It is fitting that the most deadly substance known is named after Pluto, Greek god of the underworld and ruler of the dead.

What is the future of nuclear power in general and the fast-breeder in particular? Among the industrial nations, France appears most committed to the development of nuclear power and the breeder.[37] The nuclear tide in Europe appears to be ebbing, however. Among the forces that are giving pause to nuclear programs both here and abroad are the lack of permanent disposal methods for nuclear wastes,[38] accidents at existing nuclear facilities,[39] concern about the health hazards of even minute quantities of radiation,[40] serious doubts about the economic health of the nuclear industry,[41] uncertainty regarding the commercial viability of nuclear power,[42] reservations and dissent over safety even within the industry,[43] and a general public distrust of nuclear power.[44] In the U.S., the largest and most influential industrial nation, government projections of the future role of nuclear power have been reduced to less than one-third those of a few years ago,[45] and commercial orders for new nuclear power plants have fallen to levels that cannot long sustain the industry[46] (Figure 4.2).

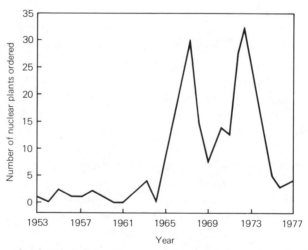

4.2 Commercial orders for new nuclear power plants over the past two decades. Orders have fallen sharply from a peak around 1973. (after J. Vinocur, "Outlook Held Bleak for Manufacturers of Atomic Reactors," *New York Times*, 21 September 1977, p. A1)

Likewise, the U.S. fast-breeder program is in a state of flux. For the moment the program has been "indefinitely postponed" as part of the Carter administration's energy program. The prototype breeder under development in Tennessee has been suspended, and all construction, commercialization, and licensing efforts for the U.S. breeder program have been ended. The timing of these decisions is extraordinarily significant, for if the fast-breeder reactor is to provide commercial power in time to compensate for the depletion of fossil fuels (i.e., before 2000), a commitment to deploy the reactors must be made at least by 1985. With the curtailment of the breeder program in the U.S. by an administration that will hold office at least until 1981 and perhaps to 1985, the decision may have effectively been made. Weinberg described the adoption of the breeder reactor as a "Faustian bargain"—acceptance of "inexhaustible" energy in exchange for eternal vigilance over the deadly radioactive by-products.[47] For the present, at least, it would appear that the world's largest industrial nation has declined the Faustian bargain offered by the fast-breeder reactor.

Fusion

There is only one nuclear alternative to fission, and that is fusion. In the broadest sense fusion already supplies most of our energy, for fusion is the basis of radiant energy from the sun and stars. Fusion can also be made to occur on earth, in the form of a hydrogen bomb explosion, but controlling the reaction so that it yields sustained, manageable energy is another matter.

Fusion resembles fission in that a small amount of matter is converted to enormous quantities of energy. Fusion is the reverse of fission, however, in that instead of breaking one atomic nucleus apart into two small ones, two small nuclei unite or fuse to produce a larger nucleus of a different element. The large nucleus is smaller than the original two combined, and the difference in mass is coverted to energy and released as heat. There are several possible fusion reactions, but the likeliest candidate for commercial exploitation involves two forms of hydrogen, called deuterium and tritium, which fuse to form helium gas (Figure 4.3).

Fusion has several advantages over fission. First, the fuel is relatively abundant. Deuterium can be readily extracted in large quantities from ordinary sea water. Tritium is rare in nature, but can be "bred" much like plutonium is bred from ^{238}U. To breed tritium, molten lithium is bombarded with radioactivity. Hence, designs for fusion reactors include a blanket of molten lithium surrounding the radioactive core in which fusion takes place. By this arrangement, fusion therefore does not suffer

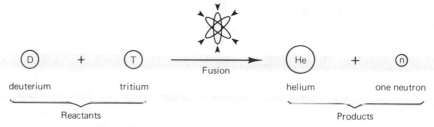

deuterium   tritium   helium   one neutron

Reactants   Products

4.3 Nuclear fusion involving two forms of hydrogen (deuterium and tritium), as would occur within a fusion reactor. The two hydrogen nuclei fuse, yielding a nucleus of helium gas and one neutron. The released neutron would enter a molten lithium blanket surrounding the reactor core to generate more tritium fuel from the lithium. The mass of the reactants is greater than the mass of the products. The "lost" mass is converted to energy and realized as heat. (modified from G. M. Crawley, *Energy* [New York: Macmillan, 1975], p. 136)

direct limitations of fuel supply inherent in fission; also, the products of fusion (mainly helium gas) are comparatively harmless. Finally, it is theoretically possible to obtain electricity directly from the fusion reaction, bypassing the many energy conversions that are now required to generate electricity.[48] The thermodynamic efficiency of such a process is estimated at 80%, which means that for a given quantity of usable power fusion would generate about half the thermal pollution of fission or fossil fuels. With "unlimited" electricity, sea water could be electrically dissociated into hydrogen and oxygen. Existing internal combustion engines could burn hydrogen with comparatively minor modifications, yielding pure water as the only product, and hydrogen could substitute directly for natural gas. Thus the development of fusion could lead to a comparatively pollution-free, energy-rich "hydrogen economy."[49]

Owing to such qualities, many energy experts consider fusion to be a "dream" source of clean, virtually unlimited energy. Controlled fusion may well remain a dream, however, for the technical problems that must first be mastered stagger even the boldest imagination. Controlling a fusion reaction is like trying to confine a hydrogen bomb explosion in a room the size of a small office without allowing the reactants to touch the walls. Materials that can withstand such temperatures and radioactivity are not yet known to science. They will have to be developed and manufactured cheaply and in quantity before fusion can become a commercial reality. But this obstacle is merely number one on a very long list.

To harness fusion, the basic task is to bring hydrogen atoms together with sufficient force to make them fuse, without expending more energy in the process than is released by the fusion reaction itself. This is no trivial task, for the hydrogen nuclei all carry the same positive charge and thus tend naturally to repel one another. Moreover, it is not enough to fuse one, nor two, nor even a billion hydrogen nuclei. To generate net power, it is essential to fuse 100,000 billion hydrogen nuclei per second within each cubic centimeter of the reactor core. So far this breakeven point, known to physicists as "Lawson's criterion," has been attained only in sophisticated research laboratories, and then only for a fraction of a second. Two main technical approaches are being pursued: magnetic confinement,[50] and laser-beam fusion.[51] *

In magnetic confinement, the two forms of hydrogen that are to be fused are heated to a temperature hotter than the surface of the sun. At this temperature the hydrogen atoms enter into a fourth state of matter, a disorganized cloud of electrons and bare atomic nuclei called plasma. The behavior of such plasma is by no means fully understood; in particular, it shivers and shakes in unpredictable waves that physicists call "plasma instabilities." These instabilities make the plasma hard to keep under control. And yet, if hot plasma were to touch the wall of the container the reactor would instantly melt. Thus the plasma must be confined to the center of an otherwise evacuated chamber, a feat that physicists hope to accomplish using magnetic force. Magnets strong enough to confine plasma must be made in part of superconducting materials that are cooled nearly to absolute zero. So, within a single machine a few meters in diameter the hottest and coldest temperatures on earth must routinely coexist. The first wall cannot be too thick or it will not let heat and radioactivity pass from the central core to the surrounding blanket of molten lithium to generate tritium fuel. Neither, however, can the first wall be too thin, for then it would wear out too fast and also pass too much heat and interfere with the operation of the supercooled magnets that confine the plasma. It is a wedding of heaven and hell, a classic paradox that may not admit to satisfactory technical solution.

In the second major approach to fusion power, laser-beam implosion, a tiny pellet of frozen hydrogen is blasted with a powerful beam of laser light. The surface of the pellet vaporizes and explodes, imploding the center and causing hydrogen nuclei there to fuse. In order to generate net energy, however, the laser beam must be aimed at the hydrogen pellet from all sides by a complicated array of perfectly aligned mirrors. Lasers strong enough to accomplish fusion are gigantic, occupying buildings that stretch for city blocks. So far, more energy has been required to power the lasers than is returned by the fusion reaction. Also, in order to generate net power several hydrogen pellets must be detonated per second.

* A third approach, electron-beam fusion, has not yet progressed beyond speculation.[52]

Thus a complex optical system will have to remain perfectly aligned while miniature hydrogen bomb blasts occur several times a second a few meters away.

Even if fusion can be controlled in the research laboratory, it is a long road to commercial application. As presently conceived, fusion reactors will have to be much larger than fast breeders (Figure 4.4) and will cost an estimated three to five times as much to build (perhaps $5–$7 billion each). It is not known how long the internal parts would last; estimates for the first wall range from 1 to 10 years.[53] If the first wall lasts less than a few years, more energy would have to be expended to build, operate, and repair the reactor than could be obtained from its operation.[54] Thus the fusion reactor could not yield net energy for industrial use. The maintenance problem alone may prove insurmountable; repairs to the intensely radioactive core of the fusion reactor will have to be carried out using robot technology that is not yet even conceived.[55] As if these problems

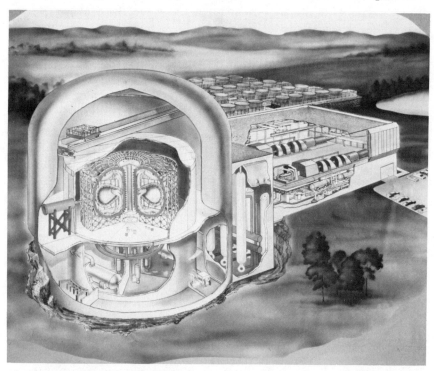

4.4 A hypothetical fusion power plant in cutaway view. The heart of the system is the array of toroidal field coils (doughnut-shaped objects in center), within which the fusion reaction would occur. The major structure housing the reactor would be made of stainless steel. The gigantic scale is suggested by the automobiles at the far right margin of the picture. (courtesy of the Information Office, Princeton University Plasma Physics Laboratory, Princeton, New Jersey. This project was funded by the Atomic Energy Commission.)

were not enough, the lithium from which the fuel would be bred is in short supply. It is not certain that there are sufficient lithium resources to meet the projected requirements of a fusion economy.[56]

Many talented and highly trained people are working to make controlled fusion a reality, and it is not inconceivable that they will eventually succeed. As Crawley has noted, however, the scientific feasibility of the breeder reactor was first demonstrated more than 30 years ago;[57] despite the investment of billions of dollars, the U.S. does not have a single commercial breeder reactor and continuation of the program is in question. There is no *a priori* reason to expect the much more difficult fusion program to fare better. The most hopeful of the fusion experts do not believe that a workable prototype of a fusion reactor will exist before 2000, and commercial deployment, if feasible, would require at least several additional decades. Fusion may have long-range possibilities, but it is given little chance to rescue industrialism from the imminent depletion of fossil fuels.

Energy Income

What, then, are the choices left to industrial economies? For the past century, the wheels of industrialism have been built and turned by depleting the earth's storehouse of fuels. When these storehouses are emptied near the end of this century, industrial society will necessarily balance its energy budget. That is, the daily expenditure of energy will of necessity precisely match the daily income of energy to the planet.

Tidal and Geothermal Energy

We have already seen (Chapter 1) that there are three major sources of energy income to the earth: tidal, geothermal, and solar energy. Tidal energy is harnessed already in parts of Europe, and could theoretically be developed further in regions such as Canada's Bay of Fundy.[58] Regions in which tidal fluxes are sufficient to generate economic power are limited to a few locations on earth, however. Geothermal energy is also tapped now, in northern California and New Zealand. Fissures through which geothermal energy can reach the surface are again confined to a few spots on the planet. Moreover, at least some geothermal installations generate as much pollution as fossil-fuel plants.[59] Both tidal and geothermal energy can provide a useful energy supplement in geographically restricted areas, but neither of these alternative energy sources can supply more than a small fraction of the total energy used today by industrial societies.

Solar Energy

In contrast to tidal and geothermal energy, solar energy is abundant throughout the middle latitudes of the planet. On the land area of the U.S. alone the sun is estimated to deliver 500–1,000 times more energy than is used by our industrial economy in an equivalent time span.[60] This energy takes many forms: heat, plant biomass, winds, hydroelectricity, and even thermal gradients in the sea that could in principle be tapped for power.[61] Likewise, the various forms of solar energy can be collected in a diversity of ways. Phototvoltaic systems, for example, convert sunlight directly to electricity, using special materials that "release" electricity when struck by light.[62] The sun's heat can also be collected directly and focused on boilers to generate steam that in turn drives the turbines of electric generators[63] (Figure 4.5).

The advantages of solar energy are many: it is abundant, clean, and effectively inexhaustible.[64] As is increasingly recognized, however, the

4.5 Artist's drawing of a 60-megawatt solar generating plant following the "power tower" concept. Such plants are designed around a large-scale, highly centralized power-generation system, and are still in the experimental stage. (after W. D. Metz, "Solar Thermal Electricity: Power Tower Dominates Research," *Science* 197 [1978], Figure 3; courtesy Electric Power Research Institute)

disadvantages are also numerous. The ultimate limit on solar energy is its diffuse nature. As we saw in Chapter 1, work can be done only when there is a temperature difference between two places. Sunlight, however, is spread evenly over large land areas. The concentration of solar power on a typical section of the U.S. is around 1,000 watts per square meter,[65] compared with a power density of 20,000–50,000 watts per square meter of surface area in an auto engine or an electric power tool.[66] Thus to do useful work, sunlight must first be concentrated enormously.

In the process of concentrating solar energy, it must be converted to another form. The efficiency of such conversions, however, is low. We saw earlier that plants convert solar radiation to chemical energy with an efficiency of 1%. The efficiency of manufactured devices for converting solar energy varies greatly, from one-millionth of a percent (windmills) to 2%–12% (solar photocells) to 60%–75% (direct-focus collectors).[67] If we include energy expenditures for the development, manufacture, operation, and maintenance of solar collectors, and the loss associated with conversion to electricity (focus collectors), the maximum realizable efficiency is probably about the same as green plants, that is, 1%. In this case the recoverable solar energy available at the surface of the U.S. is 5–10 times the current U.S. energy budget. This means that in order to power our industrial economy from the sun, somewhere between 10% and 20% of the total U.S. land area would have to be covered with solar collectors, a land area that is the equivalent of several large states.

The diffuse character of solar energy is merely one of several difficulties that are destined to limit power from the sun. As yet, for example, there is no economically feasible technology for storing electricity in large quantities after it is generated. Therefore solar power systems are useless on cloudy days and at night. Energy storage systems such as the lead–acid battery and mechanical flywheels are under development, but their realization is at least a decade away,[68] and they may never prove economical. Moreover, significant nontechnical impediments to direct solar conversion also exist, some of which we will raise later (Chapter 8). These problems have induced some to consider plant biomass as the most promising source of solar energy.[69] The use of biomass as a fuel source is still constrained by the low conversion efficiency of plants, and by competition with agricultural land and resources. Biomass may eventually supplement fossil fuels in industrial economies, but it can never replace them.

Solar technology is receiving increasing attention and research funding, and it will no doubt be used more over the next several decades. The bulk of present research, however, is concentrated on the large "power tower" concept, designed exclusively for large electric utilities and centralized power generation (Figure 4.5). A prototype commercial plant is not scheduled for completion until sometime in the 1990s.[70] The most feasible immediate application for solar energy is in heating and cooling new homes.[71] As for the exact contribution of solar energy to industrial economies, forecasts differ dramatically. In 1975, the U.S. Energy Research and Development Administration published a report predicting that by the year 2000 all the various methods of tapping the sun's energy will contribute but 6% of the total U.S. energy budget.[72] More recently, a different government agency suggested that as much as 25% of the U.S. energy budget could be provided by the sun by the year 2000.[73] The true potential contribution of sunlight probably lies nearer the low estimate. Even assuming accelerated technological development, all forms of energy income—tidal, geothermal, and solar—will probably provide no more than 10% of the projected energy requirements of industrial nations by the year 2000.

Energy Conservation

Many energy experts are increasingly convinced that alternative energy forms cannot be exploited quickly enough to replace fossil fuels. In this case, industrialism can be prolonged only by conserving existing energy supplies. Indeed, conservation of energy is the major thrust of the energy Program proposed by the Carter administration in April 1977. In theory, conservation is capable of reducing U.S. energy consumption without lowering the material standard of living. Several European nations, including Sweden, for example, enjoy per capita gross national products similar to the U.S. and yet their per capita energy expenditure is two-thirds that of the U.S.[74]

Energy conservation might be implemented by a variety of measures. For example, industrial societies currently discharge about half their energy budgets into the atmosphere as "waste" heat.[75] Theoretically, some of this heat could be reclaimed and used to heat nearby homes and businesses. Similarly, if all U.S. homes had the optimum amount of insulation, energy consumption for residential heating in 1971 would have been 42% of the actual figure.[76] A 10%–15% energy savings could in principle be realized in a variety of industries,[77] and also in transportation.

In practice, however, energy conservation is not so easily accomplished. Certain European nations get along on less energy because their societies have evolved in different directions from the U.S. Sweden, for example, has a well-developed system of public transportation that frees its citizens from dependence on the automobile.* To conserve energy in the U.S. by emulating such examples would require unthinkable capital outlays at a time when investment money is increasingly scarce. Likewise, reclaiming waste industrial heat to warm homes would require enormous amounts of money and would also require moving homes next door to power plants. In short, immense costs and ingrained institutional patterns frustrate the implementation of many possible conservation strategies.

Even if energy conservation were technically easy to implement, it is not certain that significant conservation steps could be voluntarily realized in a free capitalistic market. The U.S. automotive industry provides a case in point. In recent years, the production of large cars has set new records while the demand for small cars has withered.[79] At that same time, gasoline consumption in the U.S. has surged to unprecedented levels despite record high prices.[80] Thus the largest single market in the U.S. is apparently less sensitive to long-term energy trends than to short-term profits. The behavior of the auto industry illustrates what may be a fatal flaw in the private market economy, namely, its capacity to function successfully in the short term while ignoring or postponing the long-term consequences. The example of the auto industry raises serious questions about the potential for voluntary energy conservation under the conditions of a "free" market.

Despite the difficulties of implementing effective conservation measures, many people retain hope that here lies the potential redemption of industrialism. Proponents of conservation point out that energy use in the U.S. has actually declined in recent years. While this is true, it is also the case that the economy has suffered seriously as a consequence. Proponents of conservation suggest that American society has proved itself capable of reacting to crises almost overnight with heroic measures. During World War II, for example, the U.S. implemented extraordinary conservation measures. No automobiles were built for two wartime years, and broad personal sacrifices were willingly made on behalf of the war effort. But the present circumstance of industrial civilization is not comparable. In W.W. II, auto manufacturing was shifted to the production of tanks, trucks, planes, and ships, with a net gain of jobs. Indeed, the wartime economy was so starved for labor that government-sponsored day-

* Moreover, Sweden's economic well-being may be illusory, a lingering consequence of a more abundant past. Even with its energy efficiency the Swedish economy is in a state of disintegration, stimulated largely by the scarcity of energy and resources.[78]

care centers sprang up overnight to stimulate women to enter the work-
force. Now with the impending decline of fossil fuels industrial output
must inevitably decline, with an attendant net loss of jobs. In W.W. II,
U.S. society was united as one against a clearly defined external enemy.
Now industrialism is its own "enemy," and society is divided as to how
best to meet it. In W.W. II, Americans committed themselves to what
was perceived as a relatively brief period of sacrifice which they expected
would be followed by a materially improved life. Now the battle will last
indefinitely; what is required is a long-term change in personal values and
lifestyles, backed by the acceptance of inevitable decline in the material
standard of living. It is not clear whether a centralized democratic government can implement the necessary changes in advance.

As fossil fuels become depleted and prices rise accordingly in the
next two decades, it will become more profitable to conserve energy. The
expressed goal of the Carter administration is to reduce energy growth
from the historic annual figure of nearly 5% to 2%. Many experts consider
even this comparatively modest goal to be unreachable for at least a decade. Moreover, as we saw in the preceding chapter, the U.S. will exhaust its domestic supply of oil and natural gas by the year 2000 even if it
could immediately reduce energy growth from the past figure of 4%–5%
to 2.5%. At the very most, energy conservation can prolong industrialism
by perhaps one or two decades, and help to evolve the kind of ethic and
lifestyle appropriate to a future of comparative scarcity.

The Future of Energy

What, then, is the solution to the energy crisis? It seems increasingly possible that there may be no solution. In the short run (until 2000) we may
reasonably expect that several of the alternatives to fossil fuels discussed
in this chapter will contribute more to our energy budget. We may expect
increasing diversification and decentralization in our energy supply,[81]
depending on geography and technology. And we may expect that conservation—voluntary or otherwise—will become more widespread.

The long run is more difficult to judge, since no one can predict the
future with certainty. We can, however, speak in terms of probabilities.
From the limited vantage point of the present, it seems probable that no
single alternative to fossil fuels, nor even a particularly favorable combination of alternatives, can provide enough total energy to sustain industrialism at today's levels. The evidence summarized in this chapter

suggests that the citizens of industrial nations and their governments must soon face a new and difficult reality—a future of steadily declining energy. Moreover, as we shall discuss in the coming chapters, the "energy crisis" is only the tip of the proverbial iceberg. Beneath the surface is the prospect of depleted stores of natural resources other than energy. It is to these resources and their future that we now turn.

NOTES

1. White House Energy Staff. "Text of Fact Sheet on the President's Energy Program," *New York Times*, 21 April 1977, p. H48.
2. A. L. Hammond, W. D. Metz, and T. H. Maugh III, *Energy and the Future* (Washington, D.C.: American Association for the Advancement of Science, 1973); L. E. Swabb, "Liquid Fuels from Coal: From R & D to an Industry," *Science 199* (1978): 619–22.
3. Swabb, "Liquid Fuels from Coal," p. 619.
4. Ibid.
5. Ibid.
6. A. L. Hammond, (1976): "Coal Liquidification Plant Goes Ahead," *Science* 194 (1976): 712.
7. E. Cowan, "Search for Synthetic Fuels Delayed," *New York Times*, 7 June 1976, p. 45.
8. Ibid.; V. K. McElheny, "Synthetic Fuels in U.S. Still Held Costly," *New York Times*, 10 February 1976, p. 53.
9. W. D. Smith and E. Cowan, "Oil Dearth and Price Rise Forecast; G.A.O. Suggests Importing More Fuel," *New York Times*, 25 August 1976, p. 45.
10. A. M. Squires, "Clean Fuels from Coal Gasification," in *Energy: Use, Conservation and Supply*, edited by P. H. Abelson (Washington, D.C.: American Association for the Advancement of Science, 1974), pp. 77–82.
11. "Coal Gasification Pilot Plant," *Science News* 111 (1977): 280.
12. W. D. Smith, "U.S. Group Lists Coal Gasification Gain," *New York Times*, 17 March 1976, p. 51; V. McElheny, "Conoco Succeeds in Producing a Synthetic Gas from Lignite," *New York Times*, 1 April 1976, p. 43.
13. Hammond, Metz, and Maugh, *Energy and the Future*.
14. R. Wishart, "Industrial Energy in Transition. A Petrochemical Perspective," *Science* 199 (1978): 614–18.
15. Ibid.
16. "Oil Shale Program of U.S. in Colorado Is Suspended for Year," *New York Times*, 24 August 1976, p. 37.
17. "Oil Is Allocated after Canadian Cut," *New York Times*, 31 January 1976, p. 35.
18. Wishart, "Industrial Energy in Transition," p. 617.

19. H. E. White, *Modern College Physics*, 3rd ed. (New York: D. Van Nostrand, 1956), p. 780.
20. G. M. Crawley, *Energy* (New York: Macmillan, 1975), pp. 103–29.
21. M. A. Liebermann, "United States Uranium Resources—An Analysis of Historical Data," *Science* 192 (1976): 431–36.
22. D. Burnham, "G.A.O. Says Ford Plutonium View Perils Future of Breeder Reactor," *New York Times,* 1 February 1976, p. A18.
23. Ibid.; S. Novic, "A Troublesome Brew," *Environment* 17 (1975): 60–63.
24. B. G. Chow, "The Economic Issues of the Fast Breeder Reactor Program," *Science* 195 (1977): 551–56.
25. Ibid.
26. W. D. Metz, "European Breeders. I. France Leads the Way," *Science* 190 (1975): 1279–81.
27. W. D. Metz, "European Breeder. II. The Nuclear Parts Are Not the Problem," *Science* 191 (1976): 368–72; id., "European Breeders. III. Fuels and Fuel Cycle Are Keys to Economy," *Science* 191 (1976): 551–53.
28. G. A. Vendryes, "Superphenix. A Full-scale Breeder Reactor," *Scientific American* 236 (1977): 26–35.
29. W. D. Metz, "The Breeder: French Prototype Shut Down for Repairs," *Science* 195 (1977): 972.
30. H. Kohn, "Plutonium for Sale," *Rolling Stone* 239 (1977): 40–42.
31. D. Burnham, "A Student's Bomb Design Prompts Call for More Nuclear Safeguards," *New York Times,* 24 March 1978, p. A18.
32. D. Burnham, "Report Says U.S. Cannot Account for 2 Tons of Atom-bomb Material," *New York Times,* 6 August 1976, p. A14; D. Binder, "Accounting Sought for Atom Exports," *New York Times,* 16 December 1976, p. C13.
33. C. J. Johnson, R. R. Tidball, and R. C. Severson, "Plutonium Hazard in Respirable Dust on Surface of Soil," *Science* 193 (1976): "Geologists Find Radioactive Waste from Con Ed Building up in Hudson," *New York Times,* 18 October 1976, p. 58.
34. D. Burnham, "Safety of Breeder Reactors Questioned," *New York Times,* 16 February 1976, p. 33.
35. Ibid.
36. E. Holsendolph, "Adams Says Neglected Roadbeds Are Key Factor in Rail Accidents," *New York Times,* 28 February 1978, p. 18; "Nuclear Spill on Highway," *San Francisco Chronicle,* 12 January 1975, p. 35.
37. Vendryes, "Superphénix"; W. Sullivan, "France Leading in Shift to Nuclear Power," *New York Times,* 8 July 1976, p. 36.
38. S. Rattner, "U.S. Sees No Permanent Disposal of Nuclear Waste before 1988," *New York Times,* 16 March 1978, p. 63.
39. W. Turner, "Atom-waste Blast Contaminates Ten," *New York Times,* 31 August 1976, p. 1; W. E. Farrell, Ex-Soviet Scientist, Now in Israel, Tells of Nuclear Disaster," *New York Times,* 9 December 1976, p. 8.
40. D. Burnham, "Study of Atom Workers' Deaths Raises Questions about Radiation," *New York Times,* 25 October 1976, p. 16.

41. J. Vinocur, J. "Outlook Held Bleak for Manufacturers of Atomic Reactors," *New York Times*, 21 September 1977, p. A1; D. Shapley, "Nuclear Power Economics: Report Heats up Debate," *Science* 194 (1976): 1259.

42. Metz, "European Breeders. III"; V. McElheny, "Major Review of Nuclear Power Role Begins," *New York Times*, 23 January 1976, p. 43; "Utility Regulators Review Costs of Uranium Power," *New York Times*, 16 February 1976, p. 29; D. Burnham, "Panel Hints Nuclear Reactor Electricity May Be Costly," *New York Times*, 11 April 1978, p. 23.

43. D. Burnham, "3 Engineers Quit G.E. Reactor Division and Volunteer in Antinuclear Movement," *New York Times*, 3 February 1976, p. 12; "Indian Point Safety Head, Charging Atom Peril, Quits," *New York Times*, 10 February 1976, p. 11; "Nuclear Power in a Dead End, Hooper Says," *New York Times*, 18 February 1976, p. 45.

44. C. Hohenemser, R. Kasperon, and R. Kates, "The Distrust of Nuclear Power," *Science* 196 (1977): 25–34.

45. Vinocur, "Outlook Held Bleak"; Shapeley, "Nuclear Power Economics.

46. Ibid.; "Europe's Nuclear Turn," *New York Times*, 26 August 1976, p. 32; K. Kilborn, "British Commission Urges a Delay in Widening Use of Nuclear Power," *New York Times*, 23 September 1976, p. 1.

47. A. M. Weinberg, "Social Institutions and Nuclear Energy," *Science* 177 (1972): 27–34.

48. W. C. Gough and B. J. Eastlund, "The Prospects of Fusion Power," *Scientific American* 224 (1971): 50–64.

49. Hammond, Metz, and Maugh, *Energy and the Future.*

50. Gough and Eastlund, "The Prospects of Fusion Power."

51. J. L. Emmett, J. Nuckolls, and L. Wood, "Fusion Power by Laser Implosion," *Scientific American* 230 (1974): 24–37.

52. W. D. Metz, "Energy Research: Accelerator Builders Eager to Aid Fusion Work," *Science* 104 (1976): 308–9.

53. W. D. Metz, "Fusion Research. II. Detailed Reactor Studies Identify More Problems," *Science* 193 (1976): 38–40; J. P. Holdren, "Fusion Energy in Context: Its Fitness for the Long Term," *Science* 200 (1978): 168–80.

54. Ibid.; W. D. Metz, "Fusion Research. I. What Is the Program Buying the Country?" *Science*, 192 (1976): 1320–23; W. E. Parkins, "Engineering Limitations of Fusion Power Plants," *Science* 199 (1978): 1403–8.

55. Ibid.

56. A. L. Hammond, "Lithium: Will Short Supply Constrain Energy Technologies?" *Science* 191 (1976): 1037–38.

57. Crawley, *Energy*, p. 148.

58. E.g., A. Fisher, "Energy from the Sea . . . Waves, Tides and Currents," *Popular Science*, May 1975, pp. 69–73.

59. R. C. Axtmann, "Environmental Impact of a Geothermal Power Plant," *Science* 187 (1975): 793–803.

60. Crawley, *Energy*, p. 153.

61. W. D. Metz, "Ocean Thermal Energy: The Biggest Gamble in Solar Power,"

Science 198 (1977): 178–80; C. Zener and J. Fetkorrch, "Ocean Thermal Gradient Hydraulic Power Plant," *Science* 189 (1975): 293–94; W. D. Metz, "Wind Energy: Large and Small Systems Competing," *Science* 197 (1977): 971–73.

62. Crawley, *Energy*, p. 155; H. Kelly, "Photovoltaic Power Systems: A Tour Through the Alternatives," *Science* 199 (1978): 634–43.

63. W. D. Metz, "Solar Thermal Electricity: Power Tower Dominates Research," *Science* 197 (1977): 353–56.

64. D. Hayes, *Rays of Hope* (New York: W. W. Norton, 1977).

65. Crawley, *Energy*, p. 152.

66. A. MacKillop, "Living off the Sun," *Ecologist* 4 (1973): 260–65.

67. Ibid.

68. J. McCaull, "Storing the Sun," *Environment* 18 (1976): 9–15.

69. E. S. Lipinsky, "Fuels from Biomass: Integration with Food and Materials Systems," *Science* 199 (1978): 644–51; C. C. Burwell, "Solar Biomass: An Overview of U.S. Potential," *Science* 199 (1978): 1041–48; M. Calvin, "Solar Energy by Photosynthesis," *Science* 184 (1974): 375–81.

70. Metz, "Solar Thermal Electricity."

71. J. A. Duffie and W. A. Beckman, "Solar Heating and Cooling," *Science* 191 (1976): 143–49.

72. V. K. McElheny, "Solar Energy Future. Optimism Is Restrained," *New York Times*, 31 December 1976, p. D3.

73. G. Hill, "Over Horizon, a Solar Society Is Seen Rising," *New York Times* 3 May 1978, p. 44.

74. P. M. Boffey, "How the Swedes Live Well While Consuming Less Energy," *Science* 196 (1977): 856; L. Schipper and A. J. Lichtenberg, "Efficient Energy Use and Well-being: the Swedish Example," *Science* 194 (1976): 1001–13.

75. W. Hofele, "Energy Choices That Europe Faces: A European View of Energy," *Science* 184 (1974): 360–67.

76. E. Hirst and J. Moyers, "Efficiency of Energy Use in the United States," in *Energy: Use, Conservation and Supply*, edited by P. H. Abelson (Washington, D.C.: American Association for the Advancement of Science, 1974), pp. 13–18.

77. G. A. Lincoln, "Energy Conservation," In *Energy: Use, Conservation and Supply*, edited by P. H. Abelson (Washington, D.C.: American Association for the Advancement of Science, 1974), pp. 19–26; C. A. Berg, "Conservation in Industry," *Science* 184 (1974): 264–70.

78. J. Vincour, "Sweden's Economic Success Sours," *New York Times*, 24 March 1978, p. 1.

79. "Auto Makers Expanding Production," *New York Times*, 26 March 1976, p. 55; "May 11–20 Sales of Cars up 53.2%," *New York Times*, 26 May 1976, p. 47; "Acceleration Bypassing A.M.C.," *New York Times*, 6 April 1976, p. 47.

80. W. D. Smith, "Surge in Use of Gasoline Rasies Issue of a Possible Shortage in the Summer," *New York Times*, 11 May 1976, p. 45.

81. A. Lovins, "Energy Strategy: The Road Not Taken?" *Foreign Affairs* 55, 65–96.

Chapter 5
NATURAL RESOURCES, ECOLOGY, AND ECONOMY

Energy is the driving force behind industrialism, but it is only one element in the "industrial equation" (Figure 1.1). The wheels of industrialism are turned by energy, but the wheels are themselves fashioned from natural resources. A steady flow of natural resources is required to construct, maintain, and replace the capital machinery and other physical trappings of an industrial civilization. To assess the likely future of industrialism, then, we must assess the future of natural resources, much as we have already assessed the future of energy. We shall begin in this chapter by first developing an "ecology" of natural resources. Analyzing natural resources in an ecological context will help us to see how the flow of energy is coupled with and drives the flow of natural resources. The ecological context will show how the flow of resources is linked with the production of goods and services, the availability of jobs, money, and inflation. And ultimately, the ecology of natural resources will help us to see how the law of entropy must eventually squeeze the flow of resources to a trickle.

What Is a Natural Resource?

The dictionary tells us that a natural resource is "a capacity or material supplied by nature."[1] By this definition, energy is a natural resource since it imparts the capacity to do work and it is supplied by nature. But when we speak of a natural resource, we generally think beyond energy to such basics as land, water, forest products, and mineral deposits. Indeed, our definition of a natural resource carries us much further. Our genetic heritage, for example, although not usually considered a natural resource, in fact accords with our definition. Genes are molecules of nucleic acid, stamped indelibly by evolution with the "blueprint" of our species. Genes ultimately provide us with the capacity for everything that is human, and they are supplied by nature through natural selection. Likewise, the genetic heritage of domestic animals and agricultural plants is also a natural resource in that it confers on us the capacity to survive, and is supplied ultimately by nature. As we shall see in a subsequent chapter, there is dangerously insufficient awareness of genetic heritage as a natural resource, with the result that essential genetic variability is vanishing from our crops and domestic animals.

The Classification
of Natural Resources
Renewable vs.
Nonrenewable Resources

It has become customary to identify natural resources according to whether or not they are renewable (but for an alternative, see O. S. Owen).[2] A renewable resource is one that is constantly replenished with the passage of time. Renewable resources thus generally include those that are derived from sunlight and formed by photosynthesis, that is, all plant and animal products. A nonrenewable resource, in contrast, is not continuously regenerated, but rather represents a one-time-only "gift" to planet earth. Mineral deposits in the earth's crust, for example, are typically considered nonrenewable. As far as we know, such deposits represent a legacy of the formation of our planet eons ago from a cloud of interstellar gas.

The renewable/nonrenewable scheme has the virtue of appropriately emphasizing the impermanence of certain natural resources. The scheme is misleading, however, in basic respects. Fossil fuels, for example, are generally classified as nonrenewable; in fact we have seen that they are continuously formed by a combination of photosynthesis and slow geological processes. Likewise, metal ores are usually classified as nonrenewable, and yet new deposits are constantly formed as part of the natural cyclic exchange between the earth's molten interior and its solid crust. To designate fossil fuels and metal ores as nonrenewable denies the process by which these natural resources are continuously reformed. Fossil fuels and metal ores are nonrenewable only in the sense that they are consumed by industrialism faster than they are replaced by nature. In this case, the term "nonrenewable" tells us less about the natural resource itself than about its rate of use; indeed, a renewable resource can become nonrenewable simply by a steady increase in its consumption rate.

Seen in another perspective, however, most resources are nonrenewable, for most renewable resources depend ultimately on some nonrenewable resource for their production. Wood, for example, is renewable because trees are continuously formed anew by photosynthesis, but nonrenewable land is needed to grow trees and nonrenewable fossil fuels are needed to cut logs and mill them into useful lumber. In other words, the renewable/nonrenewable scheme can obscure interdependencies by which "nonrenewable" resources limit the production of "renewable" ones.

Primary, Secondary, and Tertiary Resources

We need some more consistent method of classifying natural resources. One such method is offered in Figure 5.1; here natural resources are classified as primary, secondary, or tertiary, according to how close they are to their original, native form.[3] We will soon see how this scheme permits a number of important insights into the ecology of natural resources.

Primary natural resources are the basic raw materials provided by nature, from which all other natural resources are ultimately derived. Land, for example, is a primary natural resource from which food and mineral ores are secondarily obtained. Primary resources also include the oceans, fresh water, sunlight, and the atmosphere. Many of these variables interact to form the climate, which as we shall see in the next chapter is one of the most basic primary natural resources. Likewise, as we have discussed, our genetic heritage is an irreplaceable primary natural resource, as is the genetic composition of domestic plants and animals.

Secondary natural resources are those that are derived directly from the primary resources by the expenditure of energy and human labor. They include raw animal, mineral, and vegetable products derived from land and sea (Figure 5.1). Secondary resources in turn provide the materials from which tertiary resources are obtained, such as wool and leather,

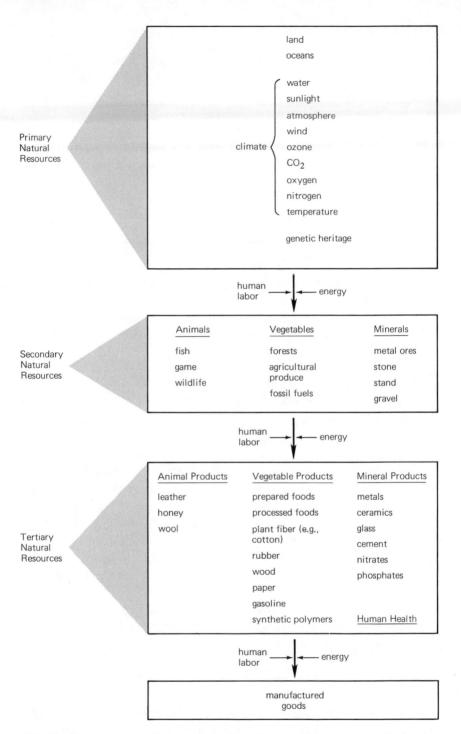

5.1 Classification of natural resources. In this scheme resources are categorized as primary, secondary, or tertiary, depending on how close they are to their original, native form.

paper, prepared foods, and smelted metal. Human health, seldom considered a natural resource, in fact corresponds to our definition of a tertiary resource. Health is prerequisite to human labor, the "capacity" by which all resources are obtained, and health depends on food, shelter, and other amenities that are supplied ultimately by nature.

The Ecology of Natural Resources

The classification scheme outlined above is particularly useful because it lends itself to an ecological interpretation of natural resources and their relationship to industrialism. We shall first develop this ecological perspective, and then explore the insights that it furnishes into such practical matters as jobs, money, and the future of industrial economies. In developing an ecological interpretation of natural resources, we will draw heavily on analogy with the food chain. As we learned in Chapter 2, the food chain is nature's way of upgrading solar energy to biomass. Plants are the primary producers in the food chain, since only they can convert solar energy to chemical energy. Plants are consumed by animals at higher trophic levels in the food chain, the herbivores, which are in turn eaten by the carnivores.

Before we can effectively draw an analogy with natural resources, three features of the food chain require emphasis. First, recall from Chapter 2 that energy from the sun—the driving force behind the food chain—is upgraded, or concentrated, as it passes through the food chain, with the attendant dissipation ("loss") of energy from one trophic level to the next. The law of entropy insists that the flow of energy through the food chain is one-way, although the nutrients that are the building blocks of life are recycled continuously. Second, recall that in the food chain organisms that occupy higher trophic levels are utterly dependent for their survival on organisms at lower levels. All life forms depend ultimately on plants, because only plants possess the metabolic capacity to "capture" solar energy by photosynthesis.

Finally, there is a third feature of the food chain that we will extend to natural resources, namely, the total biomass represented by successively higher trophic levels is progressively smaller. Thus the biomass (measured as weight) of all plants vastly exceeds the biomass of all herbivores, which in turn is greater than the biomass of all carnivores. The progressive decline in biomass at higher trophic levels forms what ecol-

ogists call an ecological pyramid[4] (Figure 5.2). The pyramid is ascribable
ultimately to the pervasive law of entropy: since energy is dissipated in
each conversion from one trophic level to the next (Chapter 2), the total
energy bound up in successively higher trophic levels is less; and less
energy means less mass, in accord with Einstein's equation (Chapter 3).

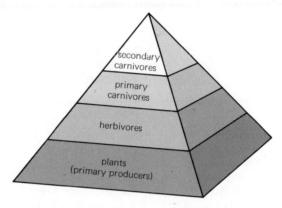

5.2   The ecological pyramid, in which organisms at progressively higher levels in the food
chain are represented in terms of the total biomass they comprise.

The Resource Chain

Exactly the same conceptual framework developed above for the
food chain can be applied to the utilization of natural resources. In the
food chain it is different organisms—plants, herbivores, carnivores—that
are the agents of energy upgrading or concentration. In the case of natural
resources, the basic human industries that extract and process the natural
resources are the agents of resource upgrading. These industries are also
organized into sequential "trophic" levels, to form what I will call the
"resource chain" (Figure 5.3).

To see how resources are upgraded by the resource chain with the
attendant dispersal of energy at each step, consider the example of copper. This metal is obtained initially from the primary resource land,
where the average crustal abundance of copper in the earth's outer man-
tle is 63 parts per million.[5] Nature has fortunately already concentrated
the copper metal, by the expenditure of energy, into deposits of copper
ore—the secondary stage of the resource. Here the metal concentration
ranges from 3,500 parts per million to as high as a few parts per hundred
in exceptionally rich ores. The ore can be made available to human use,
however, only by the expenditure and therefore dispersal of more
energy, this time by the machines of the mining and transportation indus-

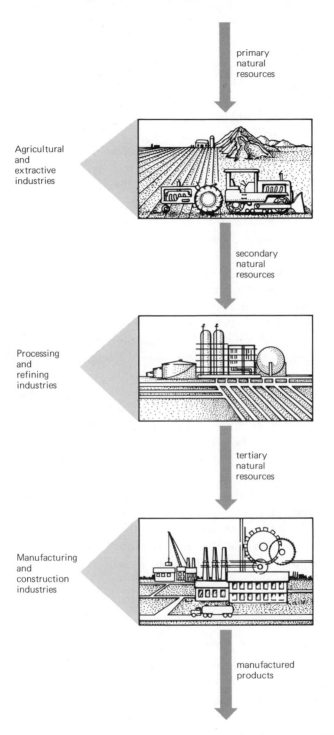

primary
natural
resources

Agricultural
and
extractive
industries

secondary
natural
resources

Processing
and
refining
industries

tertiary
natural
resources

Manufacturing
and
construction
industries

manufactured
products

5.3 The resource chain, in which human industries are organized into sequential "trophic" levels according to the class of natural resources they process. Resources are "upgraded" as they flow from one trophic level to the next.

tries. The tertiary form of copper is the purified metal (Figure 5.1) that results from smelting the copper ore. Again, however, this conversion involves the expenditure and therefore dissipation of still more energy —this time by the refining industries. Copper has then reached its most concentrated form—greater than 99 parts per hundred.

We see that whereas solar energy is concentrated in the food chain, natural resources are concentrated in the resource chain. Just as different classes of living organisms comprise the several trophic levels of the food chain, the different basic human industries can be considered as the trophic levels of the resource chain (Figure 5.3). Moreover, just as organisms at higher trophic levels in the food chain depend on lower trophic levels, the industries at higher levels in the resource chain also depend on lower ones for survival. Without the extractive industries (analogous to plants), for example, there could be no processing and refining industries (analogous to herbivores); and without these, there could be no manfacturing industries (analogous to carnivores).

The parallel between the food chain and the resource chain may be extended even further. When solar energy is upgraded in the food chain, the second law of thermodynamics insists that entropy, or randomness, must nonetheless on the average increase. Therefore, although the living matter that is produced by the food chain is extraordinarily organized, the thermodynamic price is a shrinking sun. Similarly, the inviolable law of entropy also governs the resource chain. When copper is concentrated in the resource chain, for example, it becomes less diffuse and therefore more "organized." The thermodynamic price, however, is the degradation of fossil fuels to diffuse heat. The industries that concentrate natural resources and eventually convert them to manufactured goods are energized by fossil fuels. As in the case of the food chain, the energy flow through the resource chain is one-way, although the resources (analogous to nutrients) can be recycled. Thus the availability of natural resources to society is closely tied to the availability of energy and regulated by the pervasive law of entropy.

The Resource Cycle

We noted earlier that the term "food chain" is in a sense incomplete; in fact it is organized as a cycle, in which high-quality energy is constantly fed back to upgrade low-quality solar energy (Figure 2.3). Likewise, the resource chain can be understood in terms of a cycle. In the resource cycle (Figure 5.4), high-quality or concentrated resources are fed back to upgrade low-quality or dispersed resources. To illustrate this generality, consider the example of metals. Ores are extracted from the earth and converted by the industries in discrete steps to purified metal and finally to machinery such as trucks and bulldozers. The feedback occurs when the trucks and bulldozers are then used to extract still more metal ores. Similar feedback occurs at each trophic level in the resource cycle. Thus metal machines extract ores, metal refineries smelt the ore, and metal factories turn the processed metal into yet more machines, in a cycle that is driven by the continuous degradation of fossil fuels.

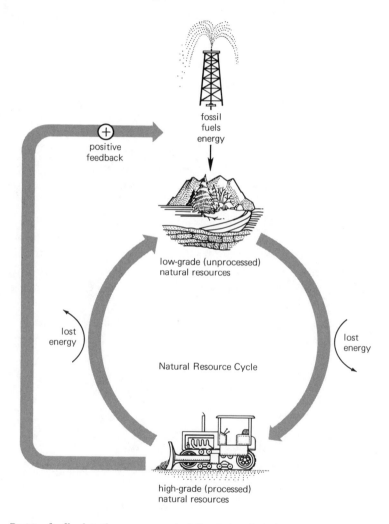

5.4 Positive feedback in the resource cycle. Manufactured products are used to extract natural resources which are used to manufacture products, etc., in a self-reinforcing process that is powered by fossil fuels. Compare this figure with the food cycle (Figures 2.1 and 2.3).

Positive Feedback
in the Resource Cycle

In Chapter 2, we noted that everything in industrial societies has grown by positive feedback that is linked to energy conversion. The explanation is to be found in a characteristic of the resource cycle. Recall that the food cycle operates by positive feedback, in which more food permits more people, who grow more food, in a self-reinforcing pattern. The resource cycle operates by the same positive feedback, necessarily coupled to energy conversion. For example, metal ores are converted in steps to bulldozers which in turn extract more metal ores, a self-reinforcing pattern that is driven by continuous energy expenditure.

Recall also from Chapter 2 that when any positive feedback process has a gain in excess of 1.0, the result is runaway exponential growth. In the case of the resource cycle, the effective gain of the positive-feedback dynamic has exceeded the magic number 1.0 for two centuries. In practical terms, this means that a single bulldozer has been able to extract more than enough metal ore in its lifetime to maintain and replace itself and the machinery that built it. Under these conditions—which require an abundant supply of metal—there is a net gain of metal available to construct still more bulldozers.

As a consequence of runaway positive feedback in the resource cycle, the past two centuries have witnessed exponential growth in the rate at which resources have been taken from the earth and fashioned into manufactured products. The physical consequence has been the accumulation of material things—roads, buildings, machines, etc. Of the many emotional consequences, one has been "future shock"—the overwhelming sensation of increasingly rapid and uncontrolled change.[6] And as we shall see next, the economic consequence has been almost unremitting growth of wealth within industrial nations.

Economics
and the Resource Cycle

As I suggested near the beginning of this chapter, an ecological perspective on natural resources can provide insights into such practical matters as jobs and money. Now we are ready to explore how entropy, energy, and economy are all related to one another through the resource cycle. Indeed, I will propose that the resource cycle is in essence equatable with the economy.

Jobs and Natural Resources

The "economy" is a complicated, dynamic, and interacting system that nearly defies definition, but certainly we can agree that one of its important products is jobs. In order to understand the relation between jobs and the resource cycle, it is helpful to continue our analogy with the food cycle. We have seen that in the food cycle, plants are the primary producers that fix solar energy so that it is available to herbivores. Similarly, agricultural and extractive industries are the "primary producers" of the resource chain in that they occupy the first trophic level and their products are essential to all other industries. Thus, as we would expect, these industries historically dominate the "childhood" of the economic life cycle of industrial nations, such as Britain in the 18th century and the U.S. in the 19th century.

The agricultural and extractive industries are fundamental to industrialism, but they could not exist without a market for their products. This market is provided by higher trophic levels in the resource cycle, beginning with the processing and refining industries. These industries typically evolve in parallel with the agricultural and extractive industries, and predominate in the "adolescence" of an industrial economy. Their products are in turn essential to the construction and manufacturing industries, which characterize the early adulthood of industrial economies. The products of these "advanced" industries in turn become essential to the operation of industries at lower levels, thus completing the cycle.

We see that the many industries that together comprise an industrial economy are tied together in a web of reciprocal dependencies, much as the lives of organisms are inseparably intertwined in the food chain. Each industry provides something that is needed by another industry, and in turn receives something in exchange. Barry Commoner has described the same dynamic as the consequence of interactions among three systems, with the

> economic system dependent on the wealth yielded by the production system, and the production system dependent on the resources provided by the ecosystem.[7]

Regardless of the particular conceptualization that is used to portray these interdependencies, the practical economic implication is identical—most jobs in an industrial economy are provided by the basic industries that comprise the resource cycle. Therefore, the availability of jobs is closely tied to the processing of natural resources, or to put it differently, employment is proportional to the rate at which the resource cycle operates. When the cycle operates rapidly to convert natural resources into manufactured goods, many jobs are available; and when the cycle slows down, unemployment results. Since the rate at which the resource cycle operates is proportional to the rate of energy degradation, the availability of jobs in an industrial economy is dependent on the availability of both natural resources and energy.

The Service Economy

As industrial economies mature, an increasing proportion of jobs is typically provided by the service industries. As implied by the name, these industries sell services rather than goods. They include recreation and entertainment, banking and finance, medicine, education, most of the "professions," and all government jobs. Because service industries come to predominate in the late stages of industrialism, some futurists have extrapolated this trend and suggested a service economy as a model for post-industrialism.[8] When we examine the ecological position of the service industries in the resource cycle, however, we are struck by their total dependence on the basic industries. To illustrate, consider one example of a service industry, education. Teachers trade their "services," the transmission of knowledge, for money. A large-scale educational enterprise requires first of all surplus agricultural productivity that frees the teacher's time from the essential activity of producing food. The teacher's job depends further on buildings, books and libraries, and a complex physical infrastructure provided ultimately by the manufacturing and construction industries. The closing of schools during winter natural-gas shortages illustrates the dependence of this service industry on the extractive and refining industries. At the level of higher education, the dependencies are even more severe. In particular, the discovery of new knowledge by research typically requires complex machinery, lengthy pretraining of research personnel, and large blocks of "free" time, all of which are made possible ultimately by "surplus" money and the products of the more basic industries.

One can perform a similar exercise with the other service industries to demonstrate their fundamental dependence on the more basic industries. To portray the dependence as a generality, we may extend our parallel with the food chain. Recall that the different trophic levels in the food chain are organized into a "pyramid," with plants at the base and carnivores on the top. The different trophic levels of the resource chain are likewise organized into a pyramid: the service industries form the apex, which rests on the broader foundation that is necessarily furnished by the more basic industries (Figure 5.5). In other words, the service industries occupy the highest "trophic" level in the resource chain, just as carnivores occupy the highest trophic level in the food cycle. The underlying reason is straightforward: just as carnivores are unable to fix solar energy, the service industries are unable to generate capital. Rather, the service industries function to circulate and redistribute capital that is generated ultimately by manufacturing and the more basic industries, much as the carnivores circulate and redistribute energy that is fixed ultimately by plants.

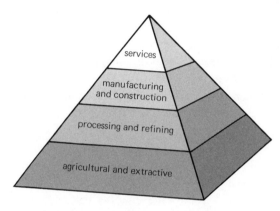

5.5 The resource pyramid, in which the basic human industries are organized in order of ascending trophic level. Compare this figure with the ecological pyramid (Figure 5.2).

If this ecological interpretation of the economy is valid, then the service industries could no more exist without the basic industries than carnivores could exist without plants. Moreover, the ecological parallel can be pushed even further; just as the total mass of carnivores is destined by natural law to be smaller than the total mass of herbivores, the total number of service jobs in an economy may necessarily be limited by the total number of jobs available in the other basic industries. That is, there may be a certain ratio of service to nonservice jobs that cannot be exceeded in a healthy, stable economy. I know of no economic theory that addresses this issue, but as we shall document more fully in later chapters, there is solid and growing empirical evidence that when industrial economies slow down, jobs in the service industries are the first to be eliminated, as would be expected from the ecological model. The matter deserves further study from economists; for if the ecological analogy is valid, it implies that a service economy, while an undeniable concomitant of industrialism at maturity, is not a feasible model for post-industrial civilization.

Money and the Resource Cycle

When we speak of the economy, money comes immediately to mind. Thus, if the economy can truly be equated with the resource cycle, it should be possible to understand money—its origin, circulation, and value—in terms of the resource cycle. Conversely, we cannot appreciate the effects of energy and resource depletion on the economy unless we understand the relation between energy and money.[9]

Paper money is an enigma. It is described as the root of all evil, even though it obviously lacks intrinsic worth—we cannot eat dollar bills, and they make a poor shelter. Money is clearly a surrogate: it has value only because it stands for something else that is generally acknowledged to have value. The "something else" was once precious metals, such as gold and silver. An inscription on some U.S. dollar bills still certifies that the bearer can exchange the note for a specified equivalent of silver. The inscription is, however, a relic of an earlier time, for now there is not nearly enough precious metal on the planet effectively to back the number of U.S. dollars in circulation.

What is it, then, that now gives paper money its generally acknowledged value? As detailed by Galbraith, the value of paper money is set largely by three variables: the total quantity of money in circulation, the velocity of its circulation, and especially the number of goods and services that can be purchased by the money.[10] To appreciate how the availability of goods and services helps set the value of money, we need only consider an extreme case. Let us imagine that, for whatever reason, goods and services are suddenly unavailable for purchase. All the money on earth cannot purchase things that are not for sale, and hence in this extreme case paper money is without value. As a rough approximation, all other things being equal, the more goods and services that are available for purchase, the greater the quantity of paper money needed to symbolize these goods and services, and the greater the value of the money.

As we have seen, goods and services comprise the output of the various stages of the resource chain or cycle. The manufacturing industries, for example, produce machines that are used by the extractive industries. In return, the extractive industries pay money to the manufacturers, money which is then returned to the extractive industries to purchase the materials for building the machines. This relationship of mutual exchange extends to individual workers within each industry, who trade their labor for money and their money for goods and services that are produced with labor. These examples illustrate an important generality, namely, that money flows in circles, in the opposite direction to goods and services[11] (Figure 5.6).

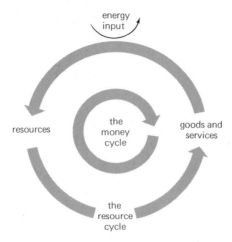

5.6 The relations among energy, resources, and money. This drawing illustrates that money flows in circles, in the opposite direction to the flow of materials through the resource cycle. That is, money is exchanged for material goods. This drawing also illustrates how both the money and the resource cycles are driven by the degradation of energy.

Because goods and services are generated by the resource cycle, and because money has value in proportion to how many goods and services are available, it follows that paper money has value in direct proportion to how fast the resource cycle operates. When the cycle operates rapidly, many goods and services are available, money is correspondingly abundant, and times are judged prosperous. When the resource cycle slows, fewer goods and services are available, the relative value of paper money declines, and times are judged harsh. Of course, the rate at which the resource cycle operates is coupled to the rate at which energy is used. Hence, as a good approximation, money is generated by energy conversion (Figure 5.6). The tangible expression of this fact is a nearly perfect linear relationship between energy conversion and Gross National Product, which is the major aggregate index of productivity in an industrial economy (Figure 5.7).

Inflation and the Resource Cycle

As we see from the above discussion, the value of paper money is not fixed. This is hardly news to the citizens of industrial nations, who have become accustomed to the decline in the purchasing power of money known as inflation. Since the early 1970s inflation has afflicted all industrial nations. Inflation is a potentially explosive political issue, and governments perpetually seek to lay it to rest with appropriate fiscal "reme-

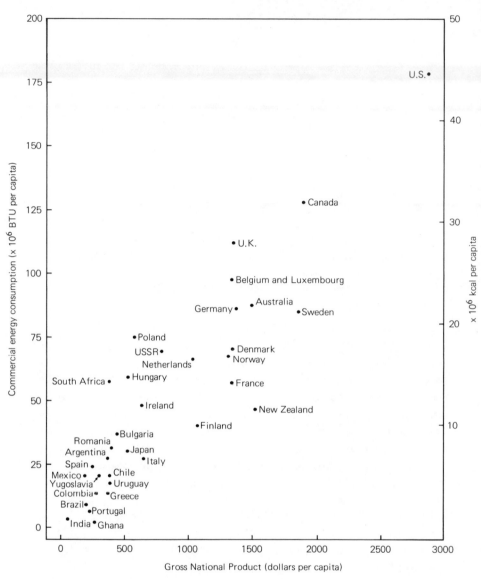

5.7 Per capita Gross National Product bears a strong relation to energy consumption. (after W. J. Chancellor and J. R. Gross, "Balancing Energy and Food Production, 1975–2000," *Science* 192 [1976]: 213–18, Figure 6)

dies." We are now in a position to interpret inflation in the ecological context of the resource cycle.[12] In so doing, we shall recognize that there is no remedy for the kind of inflation that industrial economies are now experiencing.

To see how inflation can be understood in terms of the resource cycle, imagine a hypothetical economy in which only one commodity—say, orchids—is produced. Let us assume that the resource cycle operates at a rate such that 1 million identical orchids are available for purchase at any one time, and $1 million circulate in the economy at one time. Under these higly simplified conditions, each orchid is worth exactly $1.00. Now let us suppose that a blight afflicts the orchid economy, and reduces the production rate of the resource cycle to half a million orchids per year. The $1 million in our economy are now distributed over half a million orchids. Division reveals that each orchid is now valued at $2.00, an inflationary increase of 100%. This simplified example illustrates that when the resource cycle slows down and the money in circulation remains the same, the result is inflation.

Now let us play the role of the armchair economist and tamper with the orchid economy. The blight has doubled the effective price of each orchid, i.e., lowered the purchasing power of each dollar to half its former value. Perhaps, we reason, the decline in purchasing power could be compensated by simply printing more money. It is an ancient strategy, implemented countless times throughout history. Alas, the result is always the same—more inflation. To see why, consider the consequence of doubling the money in circulation in response to the orchid blight from $1 to $2 million, without changing the productivity of the economy. Now we have $2 million distributed over half a million orchids. Division illustrates that each orchid now costs $4.00, another inflationary increase of 100%. Before long, shoppers would have to use wheelbarrows to carry enough money for the daily marketing—a circumstance that has occurred repeatedly in the evolution of most industrial nations, including the U.S.[13] Printing more money may appear at first sight a simple, attractive solution to the slowing of the resource cycle. In fact it is a siren in thin disguise, the first step in an alluring but devastating process that leads through runaway inflation to economic collapse.

We see that increasing the money in circulation does not help the orchid economy. There is another option: to reduce the money in circulation. This strategy would restore purchasing power to the dollars that remain in circulation, but because they are fewer in number the economy would be correspondingly slower. Less spending would ensue, implying fewer jobs, leading by positive feedback to still less spending. As the velocity of circulating money declines, production must eventually slow, reducing the availability of goods and services and thus lowering the value of money that remains in circulation. Exactly such an economic stagnation occurred during the Great Depression, triggered by the 1929 collapse of the stock market.[14] The orchid economy is unrealistically simplified, but it nonetheless illustrates how the value of paper money can be influenced by the amount of money in circulation as well as its velocity of circulation.

Keynesian Economics

The ecological perspective on economy that is formulated above and applied to the orchid economy invites a generalization that I believe applies fully to real economies. Namely, when the resource cycle slows down owing to constraints on the availability of energy and resources, the economy declines, and no monetary policy can reverse the decline. This view is not yet widely shared by economists; in fact, one of the most influential economic thinkers of this century, the Englishman John Maynard Keynes, taught exactly the opposite. The Keynesian solution to the Great Depression was to increase government spending through federally financed job programs and to increase consumer spending by reducing taxes. In an era of abundant energy and resources the logic of this doctrine is apparent. More spending would stimulate greater production, leading simultaneously to greater employment and more goods and services, which would stimulate yet more spending, etc. With the essential help of World War II defense spending, the Keynesian cure seemed to work. Orthodox economic theory still holds that a weakened economy can be strengthened by manipulating the supply of money.

Toward a New Economic Paradigm

To place the Keynesian solution in ecological perspective, more spending can accelerate the resource cycle. The solution is viable, however, only as long as sufficient energy and resources are available to support the acceleration. But in a time of shrinking energy and resource availability, Keynesian economic theory comprises an untenable inversion of cause and effect. As we have seen (Figure 5.6), it is energy that powers the resource cycle which in turn propels the money cycle—not the other way around. Modern economic doctrine has evolved in an era of abundant energy and resources, which has obscured this cause-and-effect relationship. As Commoner has put it, extant economic and production systems have been developed with no regard for compatibility with the ecosystem.[15]

Some economists ascribe the problems of modern industrial economies to political decisions that interfere with the operation of a free market. I believe the cause is much more fundamental: for the first time in modern economic history, industrial economies are confronted with insurmountable supply limitations on the input end of the resource cycle. The initial effect of supply limits is to increase prices, as occurred in the early 1970s when the price of oil was quadrupled. The rate at which the resource cycle operates can be sustained by paying more for natural resources, but the solution is temporary. For if merchants are to profit from their pastime, then the increased price of energy and natural resources must be passed on to the consumer. Increasing the cost of resources does not alone change the number of goods produced, however, and hence prices rise with no net increase in productivity. If the money in circulation decreases or remains the same, then each individual can purchase fewer goods, which slows the economy and encourages a recession or depression. But if the money in circulation increases, as must occur when wages go up with no corresponding increase in productivity, then more dollars are spread over the same number of goods. As we have seen from our orchid economy, the result is inflation. Odum and Odum have put the matter succinctly:

> Stimulation of the circulation of money by adding money will stimulate the flow of energy only when supplies of energy are large. Adding money when sources of energy are limited merely creates inflation.[16]

I believe it is no coincidence that the beginning of a sustained, worldwide inflation coincided with the quadrupling of oil prices. Eventually, when supply constraints can no longer be compensated by paying more for scarce resources, the resource cycle must slow. And as the orchid economy illustrates, when the resource cycle slows the combination of inflation and unemployment is the result. In the past inflation and unemployment have been considered opposite sides of the economic coin; but since the early seventies they have come together in a single package. The combination is bewildering to economic wisdom of the past, which lacks a unified theoretical framework that can cope with the consequences of energy and resource scarcity. Indeed, some economists persist in advocating old economic solutions to the world's new economic problems.[17] Other economists, however, are aware of the implications of the laws of thermodynamics,[18] and clearly recognize that a new economic

order is required[19]—one that integrates energy, ecology, and economy.[20]

The goal of this "New Economics" cannot be economic growth, for
once energy and resource production have peaked and begun to decline,
the time of economic growth is past. Neither can the goal of the New Economics be to "cure" inflation, for in a period of increasing energy and
resource scarcity, inflation is an economic expression of immutable natural law. I believe that we will have to accustom ourselves to indefinite
inflation—and a corresponding decline in the standard of living—as energy and resources are increasingly depleted. The New Economics must
be tailored to the ecosystem. Its central tenets must be consistent with
basic ecological principles, and reducible ultimately to the laws of thermodynamics. The New Economics must arise around the theme that
"Small Is Beautiful,"[21] setting as its goal the ethical distribution of increasingly scarce resources in accord with basic human needs.

As Keynes himself remarked, "the ideas of economists and political
philosophers, both when they are right and when they are wrong, are
more powerful than is commonly understood."[22] If the New Economists
can develop their discipline quickly and sell their ideas to people in
power, perhaps environmental degradation and human suffering will be
lessened in the transition to a new order.

NOTES

1. Webster's Third New International Dictionary, unabridged ed., 1971, s.v. natural resource.
2. O. S. Owen, *Natural Resource Conservation*, 2nd ed. (New York: Macmillan, 1975).
3. This scheme was derived in part from Figure 1 in S. V. Radcliffe, "World Changes and Chances: Some New Perspectives for Materials," in *Materials: Renewable and Nonrenewable Resources*, edited by P. H. Abelson and A. L. Hammond (Washington, D.C.: American Association for the Advancement of Science, 1976), pp. 24–31.
4. E. J. Kormondy, *Concepts of Ecology*, 2nd ed. (Englewood Cliffs, N.J.: Prentice-Hall, 1976).
5. Data on metallic copper concentration in various stages of the resource chain are from E. Cook, "Limits to Exploitation of Non-renewable Resources," in *Materials: Renewable and Non-Renewable Resources*, edited by P. H. Abelson and A. L. Hammond (Washington, D.C.: American Association for the Advancement of Science, 1976), p. 62.
6. A. Toffler, *Future Shock* (New York: Bantam, 1970).

7. B. Commoner, "A Reporter at Large: Energy-1," *New Yorker*, 2 February 1976, p. 38.
8. C. Clark, *Conditions of Economic Progress* (London: Macmillan, 1960). D. Bell, *The Coming of Post-Industrial Society* (New York: Basic Books, 1973).
9. H. T. Odum and E. C. Odum, *Energy Basis for Man and Nature* (New York: McGraw-Hill, 1976). See especially Chapter 6, "Energy and Money," pp. 49–59.
10. J. K. Galbraith, *Money: Whence It Came, Where It Went* (Boston: Houghton Mifflin, 1975).
11. Odum and Odum, *Energy Basis for Man and Nature*, Chapter 6.
12. Ibid.
13. Galbraith, *Money*.
14. J. K. Galbraith, *The Great Crash, 1929* (Boston: Houghton Mifflin, 1955).
15. Commoner, "A Reporter at Large."
16. Odum and Odum, *Energy Basis for Man and Nature*, p. 59.
17. P. Lewis, "Orthodox Economics for World Recovery Backed by Experts," *New York Times*, 10 June 1977, p. D1.
18. N. Georgescu-Roegen, "Energy and Economic Myths," *Southern Economic Journal* 41 (1975): 347–81.
19. L. Silk, "Some Find Keynes Policies Outmoded," *New York Times*, 21 April 1976, p. 45.
20. G. Garvey, *Energy, Ecology, Economy* (New York: W. W. Norton, 1972). See especially Chapter 3, "Energy and Ecology," pp. 61–76.
21. E. F. Schumaker, *Small Is Beautiful* (London: Blond and Briggs, 1973).
22. Quoted in J. H. Weaver and J. D. Wisman, "Smith, Marx and Malthus— Ghosts Who Haunt Our Future," *The Futurist* 12 (1978): 97.

Chapter 6
PRIMARY
NATURAL RESOURCES

In the preceding chapter we defined the resource cycle as the process by which natural resources are "upgraded" into manufactured goods. We saw further that the operation of the resource cycle is what generates jobs and money in an industrial society; thus, the future of industrialism is inextricably tied to the future of natural resources. In this chapter we will examine each of the primary natural resources—the climate, water, the oceans, air, and land—with two goals in mind. First, we shall see how industrialism depends on each of these basic resources. Second, insofar as possible we shall project the likely future of each primary resource over the critical period of the next few decades.

The Climate

We usually think of the climate as something that just happens, and not as a natural resource. But as dramatized by the recent drought that gripped North America and Europe, the climate is a most essential natural resource. The climate governs the amount of rain that falls, which in turn affects food production. The climate regulates temperature, which also strongly influences the amount of food that can be produced. Temperature dictates how much time, energy, and resources we must devote to clothing and shelter. The climate even helps to determine geography, by covering continents with icy glaciers, cool, oxygen-producing forests, or barren deserts. The climate is, in short, a critical primary natural resource that helps determine human carrying capacity, i.e., the number of people that can be supported on a given area of land. Climate is so important to human affairs that some scientists ascribe even the decline of past civilizations to sudden adverse changes in climate.[1]

The Lessons of Climatic History

Climatology, the study of the climate, has emerged from the status of a mysterious art to a quasi-predictive science only in the last two decades.[2] A major impetus for this emergence has been the development of quantitative methods for analyzing the earth's climate over the long time spans of geologic history. Most of these methods involve examining fossil records for plant or animal matter that varies predictably with temperature. When the earth's climate is warm, for example, plants flourish, producing more pollen than in cooler times. The airborne pollen is incorporated into layers of sediment on the floors of lakes and oceans, and also in the permanent ice packs of the polar regions. By sampling and analyzing cores of these sediments, climatologists can reconstruct past temperature variations in the immediate locality with reasonable accuracy. Local temperatures are in turn well correlated with global temperature, and global temperature is the best single index of past climates. Additional clues to the climate of the past are furnished by the chemical composition of sea shells, which varies with ocean temperature.[3] Likewise, coral reefs contain a hidden climatological record. When the earth warms, the poles melt, raising the sea level by well over 100 meters.[4] The increase in sea level enables corals to build their submarine cities to new heights, leaving a delicate but enduring coraline record of past climates.[5]

When methods such as these are used to reconstruct the earth's climates of the distant past, an astonishing regularity is revealed (Figure 6.1). Periods of relative global warmth have alternated with cooler periods on a precise schedule of about 20,000 years. Superimposed on this minifluctuation is an even larger rhythm, consisting of deep troughs in the temperature curve about every 90,000 years. Each of these cold troughs corresponds to a major ice age; the warm peaks that separate the ice ages are termed interstades.

The main lesson of climatic history, then, is that the climate is not constant; rather, it fluctuates on a precise schedule involving alternate periods of warmth and cold. During the warm interstades, glaciers recede toward the poles and the sea level increases as the ice melts. During the ice ages, the glaciers expand toward the equator, blanketing much of the northern and southern hemispheres with thick sheets of ice, and the level

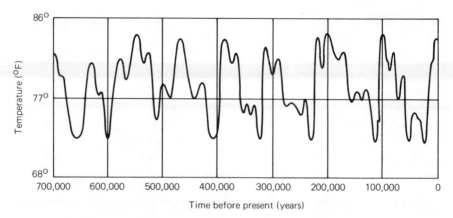

6.1 Fluctuation of temperature over geologic time. Note the regular rhythm of climatic change. Note also that global temperature of the present has reached one of its periodic peaks, perhaps in preparation for a subsequent decline. (taken from T. Alexander, "Ominous Changes in the World's Weather," *Fortune*, February 1974. Courtesy of Fortune Magazine/Parios Studios)

of the seas declines. The most recent cold period, the Wisconsin Ice Age of about 18,000 years ago, covered all of Canada and much of the northern United States with thick ice sheets.[6]

What Causes Climatic Fluctuations?

The causes of the climate are many and by no means perfectly understood. Recent evidence, however, provides a compelling and simple physical explanation for systematic climate variation.[7] It seems that our planet "wobbles" slowly on its rotational axis. When the earth wobbles, its tilt angle varies. The tilt angle in turn determines how much solar radiation enters the upper reaches of the atmosphere. This "insolation," as it is termed by climatologists, regulates the global temperature and hence the global distribution of ice sheets. For the past 500,000 years, which is as far back as data are available, the detailed harmonics of the earth's wobble have precisely paralleled and presumably caused the climate rhythm.

Why the earth wobbles is not known. Perhaps an ancient collision with a gigantic asteroid is to blame, or some quirk of our planet's early formation, hidden now in the misty reaches of time. Moreover, a wobbly earth is probably not the only contributor to climatic fluctuation. There is a long and controversial literature dealing with variation in the sun's activity as a possible short-term influence on climate.[8] In recent years it has even been plausibly suggested that ice ages may be triggered by the passage of our planet through the spiral arms of our galaxy.[9]

But for our purposes here the exact cause of weather variation is secondary. The important point is that for the past thousand millennia the climate has varied on a predictable schedule, caused by physical forces that are likely to continue into the indefinite future. As surely as the sun will rise tomorrow, the forces that underlie past climate fluctuations will be with us tomorrow.

Recent Climatic Trends

The earth is now experiencing one of the periodic temperature peaks that occur with clocklike regularity every 90,000 years (Figure 6.1). This fact alone would suggest that we are headed for a cooling trend. It is the last thousand years, however, that provides the best clues about our immediate climatic fate (Figure 6.2). Not only are these data the most precise; they are also documented in recorded history. Toward the end of this most recent millennium, that is, in the past two decades, the northern hemisphere has experienced the steepest continuous temperature drop ever recorded.[10] In the past decade alone, scientists have seen the polar ice sheets expand, increasing the earth's reflectance of solar energy back into space (the albedo) by some 12%.[11] The winter of 1978 was one of the coldest in recorded history; snow covered some 74% of North America during parts of February 1978, the most area covered during 12 years of satellite monitoring.[12] Once the albedo begins to increase, positive feedback can take over; the spread of ice sheets reflects more solar energy into space, resulting in lower temperatures, and still further expansion of the ice sheets.

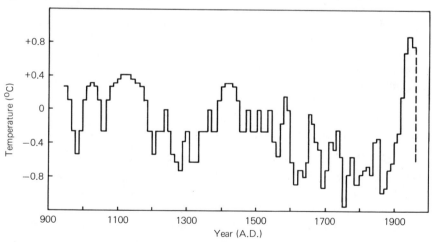

6.2 Temperature fluctuation over time in Iceland, considered a good indicator of global temperature. Note the steep decline in recent years. (after R. A. Bryson, "The Lessons of Climatic History," *Ecologist* 6 [1976]: 205–11)

The Climate of the Future

In the recent past, global coolings of the kind we are now experiencing have never reversed in fewer than 70 years, and they have lasted for as long as two centuries. Thus, barring other influences, we would appear to be headed for at least several decades of cooler weather. If the record of geological time is any indication, we may even be on the brink of a new ice age. Climatologists are reluctant to predict such major events, partially because scientists tend by training and inclination to be conservative and partially because climatology is still in its infancy. In addition, a major new variable has been added to those that affect the climate—industrialism. The large-scale combustion of fossil fuels adds particulate matter to the atmosphere, increasing the albedo and reinforcing cooling trends. On the other hand, fossil fuel combustion warms the atmosphere directly and also releases vast quantities of carbon dioxide which, as we shall discuss soon, raises atmospheric temperatures by the "greenhouse" effect.[13]

Climatologists are not certain how this ungainly mix will influence the weather of our immediate future. Indeed, some climatologists believe that the earth is warming up, based on increased temperature readings in the southern hemisphere.[14] Climatologists are agreed, however, on one point: a major change in global climate is occurring. We know from climatic history that climate changes are characterized by extreme weather instability, especially in the middle latitudes where most of the earth's population is concentrated. In the past several millennia, cooling trends such as the earth is now experiencing have never lasted less than 40 years; nor has the climate returned to its original state in less than 70 years.[15] A European meeting of climatologists in 1974 reached the consensus that

> a new climatic pattern is now emerging. There is a growing consensus that the change will persist for several decades, and that the current . . . [human] . . . food producing system cannot easily adjust. It is also expected that the climate will become more variable than in recent decades.[16]

The consensus seems validated by recent events. In this decade the monsoon rains have failed twice; drought has caused massive famine in the Sahelian region of Africa; and the U.S. has experienced its worst drought in recorded history, followed by its most extreme winter. Such variability in the weather plays havoc with modern agricultural systems, and therefore the price of food. In the spring of 1978, for example, uncommonly heavy rains in the midwestern U.S. delayed planting, which in turn thoroughly disrupted the fertilizer industry.[17] As these words are written, freak spring storms have elevated the price of California vegetables to as much as twice normal. As will be seen in future chapters, the spiraling price of food may form the lever of change in our industrial civilization.

The likelihood of imminent major change in global climate has not escaped the notice of the governments of the world. In the U.S., the Central Intelligence Agency (CIA) released a report in 1976 that was somewhat less restrained than the scientific communiqué cited above. The CIA report noted the probability of grave food shortages induced by climatic variability, food shortages that would

> prompt increasingly desperate attempts on the part of powerful but hungry nations to get grain any way they could. Massive migrations, sometimes backed by force, would be a live issue, and political and economic instability would be widespread.[18]

Whether such dramatic events will unfold in our lifetime is unclear. What seems certain, however, is that we will not soon return to the climate of the recent past, which has been so favorable to agricultural productivity. It seems probable, therefore, that the next few decades will witness not only the exhaustion of fossil fuels, but also the "depletion" of yet another natural resource that is critical to industrialism, the climate.

Fresh Water

Earth is the water planet, the only world in this solar system that has abundant quantities of the liquid. Indeed, scientists believe that the abundance of water on earth is responsible for the evolution of life as we know it. As we noted in an earlier chapter, solar energy propels water through the hydrologic cycle, from sea to rain and back again to sea. Humans obtain water by tapping the hydrologic cycle at various points, including rainfall, lakes and reservoirs, rivers and streams, and ground water.

The Uses of Water

Of the many uses of water, none is more immediate than drinking. But water plays many more roles in an industrial society. Growing a single tomato, for example, may require as much as a gallon of water. As the drought that recently afflicted the U.S. and much of the world emphasized, agricultural uses of water are critical to industrial economies. The drought cost farmers in the South and West several billion dollars.[19] When farmers lose money they cannot purchase their usual supplies and machinery. The loss to the national economy is magnified an estimated threefold by this "multiplier" effect.

Water is crucial also to the major pursuit of industrialism, manufacturing. The making of a single automobile, for example, requires 100,000 gallons of water. Likewise, the extraction of minerals, metal ores, and fossil fuels requires vast quantities of fresh water. Each gallon of oil extracted from oil shales requires 1.5 gallons of fresh water.[20] Water also plays a less direct but nonetheless important role in industrialism by providing waterways for the transportation of materials essential to manufacturing. When the upper Mississippi River froze in the winter of 1976–1977, numerous factories in the Midwest were forced to close because the barges that normally transport their fuels could not operate.

The Depletion of the Water Resource

The degradation of water in the industrial nations of the world by pollution is common knowledge. The Great Lakes of the U.S. are dead or dying,[21] and there is not a single major river system in the U.S. or Western Europe that is not afflicted with serious pollution.[22] Eventually, nature's inevitable cycles return pollution to its source, human beings. Contamination of drinking water is increasingly common in the U.S. Some communities must drink bottled water owing to industrial pollution of their former water supply,[23] and communities on the eastern seaboard of the U.S. have discovered carcinogens in their drinking water, sometimes forcing closure of public wells.[24] The lower Mississippi River, which supplies drinking water to many Louisiana residents, is so polluted by industrial effluents that its waters cause genetic changes in bacteria,[25] implying the presence of cancer-causing chemicals.[26] The effects are not confined to bacteria; there is a significant correlation between drinking Mississippi water and human cancer.[27]

Surface water supplies most of the water needs of the U.S., but underground water is 32 times as abundant, and currently supplies one-fourth[28] to one-half[29] of the nation's water. In many localities, and especially in the agriculturally productive valleys of California, ground water is being pumped to the surface faster than it is replaced by nature. As a result, the underground water level, or "table," is dropping. A decade ago drillers struck water at 10–20 feet; now the water table has dropped to below 1,800 feet in some California localities. When the water table drops in coastal regions, salt water rushes in to take its place. Such saltwater intrusion has forced the abandonment of wells along California's coast and in the agriculturally productive Salinas Valley.

Ground water is not only subject to overuse; in addition, it is damaged increasingly by accidental and deliberate contamination. Countless cases of accidental contamination have been documented,[30] many involving the leaching of toxic chemicals from surface wastes. Since ground water typically flows much like a surface stream, localized sources of contaminants on the surface can pollute much larger regions of underground water. In one instance, a single disposal site in New York leached poisons over an area of several square miles, contaminating an estimated billion gallons of ground water.[31]

Accidental contamination of ground water is small, however, compared with deliberate contamination. As surface water becomes polluted beyond its capacity to absorb more wastes, U.S. industries rely increasingly on underground "disposal" sites.[32] Municipal sewage and radioactive wastes are typically pumped into abandoned wells, and U.S. industry currently injects 400 billion gallons of wastes annually into underground disposal sites.[33] Of course, such wastes are not disposed by injecting them into the earth; they are merely stored. How and when they will return is unknown, but the eventual contamination of drinking and irrigation water seems inevitable.

The Future of Water

For the immediate future, at least, there is little prospect of reducing the degradation of the water resource. Hazardous chemicals are indispensible to the nation's industry and therefore economy,[34] and they increasingly find their way into our waters. As Zwick and Benstock point out in *Water Wasteland:*

> The situation is likely to get worse. An estimated 500 new chemicals are
> produced each year and introduced into manufacturing processes with-
> out public information concerning the extent of their dispersal through-
> out the environment, or of the dangers of that dispersal.[35]

Federal regulatory agencies are not effectively dealing with the pollution of our waters; of the approximately 15,000 potentially toxic chemicals and pollutants used and produced by industry, for example, the Environmental Protection Agency has set standards for only 16, and lacks the capacity to enforce even these.[36]

Of course, water is never destroyed or "used up." After it is polluted or used, it reenters the hydrologic cycle, to be recirculated once again. The hydrologic cycle operates, however, at a constant rate; it takes the average droplet of water 27 days to travel the complete cycle. In contrast, the use of water has increased exponentially in industrial societies such as ours. Water, like fossil fuels, is therefore being used faster than nature can replace it. Water is one of those natural resources that is rapidly converting from "renewable" to "nonrenewable," owing to a steady increase in its use rate (Chapter 5). A recent U.N. report concludes that worldwide shortages of fresh water loom on the horizon. For the peoples of many parts of the world, such shortages are already a reality. [37]

The Oceans

Three-fourths of the surface of our planet is covered by the oceans. Its fisheries are an important source of food; its minerals and fuels increasingly invite large-scale exploitation; and its algae photosynthesize an estimated 28% of the oxygen that we breathe. The oceans seem infinite, and industrialism has indeed treated them as infinite reservoirs for unwanted wastes. The oceans have even been seriously proposed as a "permanent" dumping ground for radioactive wastes. [38] When polluted rivers flow seaward, they carry their contaminants with them, but when water is evaporated from the seas to form rain, the contaminants are left behind. Every pollutant that can be carried by water eventually reaches the ocean, and those that do not decompose accumulate and become more concentrated with time.

Scientific studies of ocean pollution have been conducted mainly in the last decade. Some of these studies have received wide publicity; thus the run-off of pesticides such as DDT, and their subsequent appearance in oceanic food chains, is well known. Recent studies have disclosed a new trend, consisting of a broad pattern of contamination of ocean waters by hydrocarbons—the by-products of fossil fuel use[39] (Figure 6.3). Accidental oil spills, although a significant form of pollution, comprise less than 10% of this hydrocarbon contamination; the vast majority is caused by routine ship traffic. The potential effects of these foreign substances on marine life—including the phytoplankton that produce much of the atmospheric oxygen we breathe—is unknown. Oceanographers have also recently discovered radioactive materials (plutonium and cesium) that have leached from nuclear power plants or leaked from "sealed" drums of radioactive wastes dumped into the oceans off U.S. coasts. [40] Likewise, sewage and toxic chemicals that have been dumped directly into the oceans or have run into them in rivers have polluted beaches, contaminated sea foods, and destroyed all life in vast tracts of ocean.[41] A half-million pounds of the deadly polychlorinated biphenyls (PCBs) have been dumped by industry into the Hudson River; the General Electric Company alone is responsible for nearly a fifth of the total.[42] According to a government

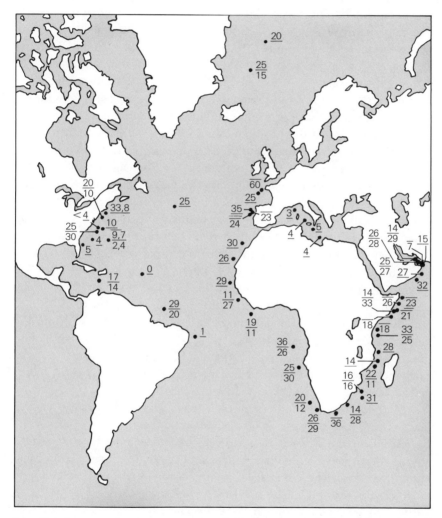

6.3 Global pattern of ocean contamination by hydrocarbons—the by-products of fossil-fuel use. Black dots indicate sampling positions. The fraction next to each dot shows the relative percentage of the hydrocarbon concentration at the surface (numerator) and at a depth of 10 meters (denominator). (after R. A. Brown and H. L. Huffman, Jr., "Hydrocarbons in Open Ocean Waters," *Science* 191 [1976]: 847–49, Figure 1)

report, these PCBs are moving downstream toward the Atlantic at a rate of 5,000–10,000 pounds per year.[43] Commercial fishing in the river is banned, but now the PCBs are appearing in seafood marketed on the East Coast.[44]

In a similar incident, Kepone—an extremely poisonous and persistent pesticide—has been discharged in large quantities into Virginia's James River, largely by the Allied Chemical Company.[45] The river has also been closed to fishing, and the Environmental Protection Agency has found evidence of Kepone contamination in the Chesapeake Bay, into which the James River empties.[46] Allied Chemical was fined $13 million for its part in what has been described as the "greatest environmental disaster of the decade"[47]—the largest settlement of its kind, but still a small fraction of the actual cost of the contamination. Once again, government regulatory agencies have proved impotent in protecting the environment and therefore people from the poisonous effluents of industrialism.[48]

One of the extraordinary and perhaps unexpected features of such incidents is the persistence of the pollution. The Shenandoah and Holston rivers of Virginia, which also drain into the Chesapeake Bay, have been found to be heavily contaminated with mercury discharged by the Du-Pont company. Upstream from the company's Waynesboro, Virginia, plant the mercury concentration in river bottom sediment is less than 1 part per million; downstream it exceeds 240 parts per million. The remarkable aspect of the case is that it has been 27 years since mercury was used in any manufacturing process at the Waynesboro plant. Most of the mercury contamination occurred in the decade from 1930 to 1940; it is still there, and may be for several decades—until it is washed into the Chesapeake to contaminate the ocean. This example[49] illustrates the longevity of pollution; even if all contamination were to cease today, we might well have to live with the consequences for decades, generations, and in the case of radioactive pollution, centuries.

With the renewed interest in the sea as a source of natural resources such as minerals and fuels, it seems certain that an increased burden will be placed on the oceans. It is simply not known how long the ocean resource can bear such abuse without an irreversible breakdown of life-support systems. The Great Lakes seem enormous and yet Lake Ontario is largely devoid of life. The seas are the acknowledged source of life on earth. Our own blood still carries the same concentration of salts as in ocean waters, a reminder of our evolutionary debt to the seas. If and when we kill the oceans with our wastes, they will die silently. But like the pounding of the surf, the repercussions will echo for the remainder of our planet's history.

The Atmosphere

Encapsulating our planet as it hurtles through space and time is a thin envelope of gas, the atmosphere. Beyond our immediate need to breathe air, we seldom consider the many roles of the atmosphere in our lives. And yet these roles are crucial to the survival of industrialism. First, the atmosphere filters out harmful solar radiation, so that when sunlight reaches earth it is relatively harmless to people and beneficial to plants. Second, the atmosphere contains not only our oxygen, but also the carbon dioxide that plants breathe. Third, the atmosphere holds and circulates moisture and heat, and thus contributes to the climate. Fourth, the atmosphere acts as a receptacle for the many effluents of industrialism, including especially particulate matter, invisible oxides, and waste heat. Finally, the atmosphere gives the sky its blue, the stars their shimmer, and conducts sound waves so that we may hear.

So vast is the sky that it has seemed, like the seas, an infinite resource. And yet even the sky has limits, announced in this century by killer smogs in London and Los Angeles and the widespread use of gas masks in Tokyo. More recently, the widely publicized threat to the ozone layer and the consequent banning of aerosol sprays has raised public awareness of the fragility of our atmosphere. Several kinds of pollution afflict the atmosphere, including "natural" pollution from forest fires and volcanic eruptions, the "London-type" sulfurous smog that has been known for centuries, and photochemical smog, discovered only two decades ago.[50] By far the largest contributor to air pollution is the direct combustion of fossil fuels. Among its documented effects—damage to human health; the despoilation of buildings, statues, and archaeological treasures by grime deposits and acid etching; the destruction of entire animal species by acid rains[51]—air pollution has three effects that are not so immediately obvious. These are impairment of plant growth, alteration of the climate, and the widespread dispersal in the air of cancer-causing agents such as asbestos.

Air Pollution and Plant Productivity

The Los Angeles Basin furnishes an excellent outdoor laboratory for studying the impact of air pollution. Here an unusual combination of geography, climate, and industrial activity collaborate to cause one of the worst air-pollution problems in the U.S. An unexpected consequence of such pollution has been the widespread damage to forests in and around

L.A. Such destruction has not only eliminated an important recreational area, but also removed a small but significant source of atmospheric oxygen.

Air pollution is unselective in its destruction of plants, also impairing the growth of agricultural crops. The yield for some fruit crops in unfiltered L.A. air, for example, is half the yield in the same air after it is filtered through charcoal.[52] In Fresno, California, the reduction of the cotton crop by one-third over the past few years is ascribed to air pollution.[53] The problem is exacerbated by the fact that 13% of all U.S. farmland falls within urban regions that are the source of air pollution. This urban farmland is among the most fertile in the nation, producing 24% of all farm income, 60% of all vegetables, 43% of all fruits and nuts, and 17% of all corn.[54]

Air pollution is by no means confined to cities, however. In sections of Europe, the eastern and midwestern U.S., and California, air pollution blankets the countryside for tens to hundreds of miles from its source.[55] California's fertile Salinas Valley—the source of most of the nation's lettuce—frequently fills to its brim with smog from upwind San José. In sections of the Midwest and Northeast, the high sulfur content of air downwind from industrial centers has caused acidic rains that destroy plant and animal life.[56] Indeed, a veil of smog is said to encircle the entire planet.[57] When air pollution reduces visibility, it is considered inconvenient; when it impairs health, unfortunate; but when air pollution reduces agricultural productivity, it retards the food cycle and strikes at the heart of the life-support system that sustains industrialism. The effect is manifest economically; air pollution causes annual crop losses in the U.S. alone of half a billion dollars.[58]

Carbon Dioxide
and the Greenhouse Effect

A second unavoidable consequence of industrialism is the release of vast quantities of carbon dioxide (CO_2) into the air by the oxidation of fossil fuels (Chapter 1). CO_2 in the atmosphere in turn captures solar energy that would otherwise bounce back into space, thus increasing global temperature by the so-called greenhouse effect.[59] CO_2 is a "trace" gas that is present naturally in the atmosphere in minute amounts. Before the Industrial Revolution the CO_2 concentration was 290 parts per million (ppm); now it exceeds 330 ppm, and is rising at an annual rate of somewhat over 1 ppm[60] (Figure 6.4). Not only CO_2, but other trace gases as well, have accumulated in significant quantities owing to industrialism.[61]

If present trends continue, which is certain if coal is used in place of dwindling oil and natural gas, then by the year 2020 the CO_2 content of our atmosphere will have doubled.[62]

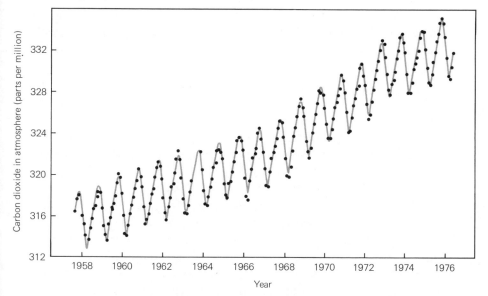

6.4 Carbon dioxide content of the atmosphere measured at the Mauna Loa Observatory on the island of Hawaii. The regular oscillations are seasonal, caused by photosynthetic "pulses." The general upward trend is caused in part by the oxidation of fossil fuels. (from G. M. Woodwell, "The Carbon Dioxide Question," *Scientific American* 238 [1978]: 34–43)

The exact relationship between CO_2 concentration and temperature is unknown, and can be estimated only with models whose parameters are uncertain. According to one calculation, however, accumulation of CO_2 alone (excluding other trace gases) could raise the average global temperature by 0.5°–0.75° Centigrade by the year 2000.[63] Yet another calculation suggests that a doubled CO_2 level, as will occur by 2020 if coal is widely used, would increase mean global temperature by 2°–3°C, with a powerful amplification in polar regions yielding a local warming of 8°–10°C.[64] These may sound like small increases in temperature; in fact major climatic changes have been accompanied by temperature changes of only 1.0°C.[65]

Climate changes of the kind that may result from heightened CO_2 levels could have a devastating impact on our civilization. Global warming, for example, would increase the world's arid lands at the expense of agricultural land. Moreover, a global increase in temperature could melt the polar ice caps, conceivably raising sea levels by 20–100 meters. Much of the world's agricultural land is located on fertile river deltas that are less than 100 meters above present sea level. Thus, among other effects, increasing CO_2 levels could drastically reduce world agricultural productivity. Because there are so many unknowns in the relationships among carbon dioxide, temperature, and climate, humanity will probably have no advance warning about such calamities. Once they are clearly evident, it may be too late to alter their course. Indeed, some scientists believe that air pollution by human activity has already significantly changed the climate.[66]

Asbestos: Trouble in the Air

Air pollution can strike in unexpected ways. One would hardly anticipate, for example, that building roads, school playgrounds, and park recreation areas could cause a long-term cancer threat. At a large quarry near Rockville, Maryland, however, crushed stone containing dangerous concentrations of asbestos fibers has been produced for years and used instead of asphalt for surfacing.[67] Sampling of ambient air 10–100 meters from roads surfaced with the crushed stone revealed asbestos concentrations 1,000 times greater than an average from 49 American cities in an earlier study.[68] Inhalation of asbestos can lead to cancer after a latency period of 20–40 years.

The problem of asbestos contamination of air may not be confined to Maryland. The mineralogy of vast regions of the U.S., including the heavily populated Northeast and West, suggests that many of the rock quarries in them may also be producing asbestos-containing stone (Figure 6.5). Without fossil fuels the asbestos-bearing rock would of course have laid undisturbed in the earth, but industrialism has both permitted and required the utilization of the resource, leading to yet a new, possibly widespread and dangerous form of air pollution.

Land

The one-quarter of the earth's surface that is land is the source of most secondary natural resources. Seventy-five percent of all human food comes from arable lands. Only 11%–15% of the earth's land surface is considered arable, but from one nation to the next the figure varies. Saudi Arabia, although rich in oil, has virtually no arable land; in China the arable land is 15% of the total; and in the U.S., 25%. Worldwide, an es-

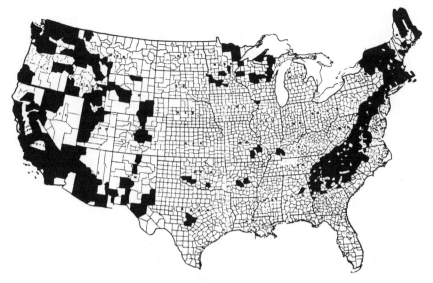

6.5 Map of the United States showing areas in which quarries may produce asbestos-bearing rock. (prepared by the Environmental Defense Fund from information derived by the Mining Enforcement and Safety Administration from reports by Batelle / Columbus and the U.S. Geological Survey as it appeared in L. J. Carter, "Asbestos: Trouble in the Air From Maryland Rock Quarry," *Science* 197 [1977]: 237–40)

timated half of all arable land is in active use.[69] In industrial countries such as the U.S., however, the pattern is strikingly different; here 80%–90% of all arable land is already under cultivation, reflecting the availability of prerequisite energy and capital.

The Expansion of Arable Lands

Since only half the world's arable lands are in use at present, some authors have expressed the belief that this resource can be expanded significantly in the coming decades. The possibility is probably illusory, for land follows the same depletion dynamic as fossil fuels. That is, the best land is developed first and what remains is exponentially more difficult and expensive to bring under cultivation. Developing more arable land worldwide would require draining swamps, grading steep hillsides, constructing massive irrigation works, building access roads, and so on. In the past two decades, when energy has been cheap and a growth ethic has prevailed, the annual rate of increase in worldwide arable land under cultivation has been only 0.15%,[70] corresponding to a doubling time of several hundred years. It is hard to see how this rate can increase much now that energy availability is declining. Even if the amount of land under cultivation could be increased in the coming decades, virtually all studies on the matter conclude that the land resources of the world are probably near the saturation point, [71] in that an increase in available land cannot alone produce enough food to keep pace with growing populations. In other words, to meet the increased food needs of the world's population over the next few decades will require obtaining more food from land now under cultivation. Moreover, as we see next, several forces conspire to reduce the availability of arable lands.

The Degradation of Land

As growing world populations put increasing pressure on land resources, large tracts of arable lands have been degraded by several forces. [72] The overgrazing of rangelands, for example, has helped to convert substantial areas of formerly useful lands to desert. In recent years, the Sahara Desert has expanded southward, adding an additional 250,000 square miles of formerly fertile land to its barren wastes. According to a recent U.N. report, such "desertification" caused by the overuse of land has now afflicted nearly 7% of the earth's surface, a land area larger than the United States. [73]

In the industrial nations, substantial amounts of arable land are lost each year to a different force, urbanization. In the U.S. alone, the cumulative loss of arable land to urbanization amounts to 45 million acres, greater than the area of Nebraska and nearly 10% of the estimated arable land in the U.S. [74] Likewise, during the last 200 years at least one-third of the topsoil on U.S. croplands has been lost to erosion, largely to water (75%), but also to wind (25%). [75] Industrial agricultural techniques are responsible for much of this loss; chemical fertilizers, for example, do not replenish organic matter that holds moisture and prevents wind erosion. Moreover, as industrial nations exhaust their energy supplies, energy needs compete increasingly for agricultural land. In the U.S., for example, strip mining disturbs an estimated 153,000 acres each year. [76] As oil and natural gas are exhausted in the coming decades, the pressure from strip mining is bound to increase.

Another source of land degradation comes from salinization. All irrigation water contains some minerals, but when ground water is overused and salt water intrudes, the mineral content of irrigation water increases dramatically. When these waters are used on cropland, the salts are left behind in the soil. [77] As pressures on the water resource increase, salinization may be expected to take an increasingly large toll on the world's present arable land. Today, nearly a third of Iraq glistens like fields of freshly fallen snow as a result of salinization. In Pakistan, 16% of the irrigated land has been partly or wholly destroyed. Likewise, salinization has ruined vast tracts of agricultural land in Mexico and the U.S. Salinization of cropland was at least partly responsible for the collapse of some ancient civilizations.[78]

Land and the Future

The consensus of informed judgment is that the land under cultivation worldwide can be increased slightly in the coming decades, but only at a disproportionate cost in energy and capital. In the nonindustrial nations, however, where food problems are most acute, shortages of energy and capital are most severe. Moreover, it is not enough simply to increase the land under cultivation; rather, the rate of increase must be greater than the rate at which arable land is simultaneously lost to degrading forces. Even when net arable land has increased, as in Brazil, it has not benefited many people. Instead, the resulting increase in agricultural productivity has been exported to richer nations to generate capital for industrialization.[79] It is safe to conclude that increases in the world's arable land resources cannot play a significant role in feeding growing populations over the next few decades.

The Future of Primary Natural Resources

In their use of primary natural resources, industrial societies are faced with a classic dilemma. As documented in this chapter, primary natural resources are being depleted at an alarming rate, in the sense that environmental degradation is reducing their quality and quantity. There is no question that despoilation of the environment in the name of economic growth impairs human health, the climate, agricultural productivity, the "quality of life," and ultimately the economy. The dilemma lies in the fact that any significant corrective action that can be imagined entails disruption that is even more immediate and—in the eyes of many—more severe. For example, to reduce fuel consumption voluntarily would reduce air pollution and slow the accumulation of CO_2 in the atmosphere. But as we have seen (Chapter 5), the economy is driven by energy; reducing fuel use would also slow the economy and reduce the material standard of living.[80] Likewise, reducing the construction of roads with asbestos-containing rock would lessen the threat of cancer, but in the absence of economically feasible alternatives, would limit road construction, transportation, and economic growth.

In short, there are costs to environmental degradation, but there are also results that are perceived as beneficial. Until the evident costs outweigh the perceived benefits, the depletion of primary resources is likely to continue. Clean air and water standards have been significantly relaxed in the economically troubled Northeast and Midwest U.S.,[81] and in some of the industrial regions of Europe such standards never existed. With regard to the costs, we can only hope that environmental degradation becomes uneconomic before it becomes irreversible. Perceived benefits can be expected to decrease only through a restructuring of human values, such that human needs are equated with environmental needs.

At the very least, it is reasonable to conclude that the primary natural resources, and especially arable land, will not increase in quantity or quality in the coming decades. This fact—together with the depletion of energy that we have discussed, and the exhaustion of secondary natural resources that we will consider next—is a necessary and sufficient condition for the thesis I will develop in coming chapters.

NOTES

1. R. A. Bryson, H. H. Lamb, and D. L. Donley, "Drought and the Decline of the Mycenae," *Antiquity* 48 (1974): 46–50.

2. S. H. Schneider (with Lynne Mesirow), *The Genesis Strategy: Climate and Global Survival* (New York: Plenum, 1976); R. A. Bryson and T. J. Murray, *Climates of Hunger: Mankind and the World's Changing Weather* (Madison, Wisc.: University of Wisconsin Press, 1977).

3. C. Emiliani, "Pleistocene Temperatures," *Journal of Geology* 63 (1955): 538–78.

4. See N. A. Morner and A. Dreimanis, "The Erie Interstade," *Memoirs of the Geological Society of America* 136 (1973): 107–34, and A. Dreimanis and A. Raukas, "Did Middle Wisconsin, Middle Weichselian, and Their Equivalents Represent an Interglacial or an Interstadial Complex in the Northern Hemisphere?" in *Quarternary Studies*, edited by R. P. Suggate and M. M. Cresswell (Wellington, N.Z.: Royal Society of New Zealand, 1975), and the references cited therein.

5. See J. D. Hays, J. Imbrie, and N. J. Shackelton, "Variation in the Earth's

Orbit: Pacemaker of the Ice Ages," *Science* 194 (1976): 1121–32, for a review of this methodology.

6. T. Alexander, "Ominous Changes in the World's Weather," *Fortune*, February 1974, p. 91.

7. Hays et al., "Variation in the Earth's Orbit."

8. J. M. Wilcox, "Solar Structure and Terrestrial Weather," *Science* 192 (1976): 745–48; W. W. Kellog and S. H. Schneider, "Climate Stabilization: For Better or for Worse," *Science* 186 (1974): 1163–72.

9. "How to Trigger an Ice Age," *Science News* 113 (1978): 148.

10. R. A. Bryson, "The Lessons of Climatic History," *Ecologist* 6 (1976): 205–11.

11. G. J. Kukla and H. J. Kukla, "Increased Surface Albedo in the Northern Hemisphere," *Science* 183 (1974): 709–13.

12. "February Sets Snow Cover Record," *Science News* 113 (1978): 148.

13. P. E. Damon and S. M. Kunen, "Global Cooling?" *Science* 193 (1976): 447–53.

14. Ibid.

15. Bryson, "The Lessons of Climatic History."

16. Ibid., p. 211.

17. H. J. Maidenberg, "Crop Planting Delays Disrupt Fertilizer Industry," *New York Times*, 5 April 1978, p. 47.

18. "CIA Weather Study Cites Global Crisis," *New York Times*, 1 May 1976, p. 2.

19. B. D. Ayres, Jr., "Drought in South Worst Since '54, Threatens Crops," *New York Times*, 29 June 1977, p. 1; "California Fears a $3 Billion Loss as Drought Parches Rich Farms," *New York Times*, 22 February 1977, p. 1.

20. For an excellent analysis of the dependence of energy resources on fresh water, see J. Harte and M. El-Gasseir, "Energy and Water," *Science* 199 (1978): 623–42.

21. "Great Lakes Study Finds Cleaning up Could Take Decade," *New York Times*, 12 March 1976, p. 31; R. Stevero, "New York Cancels Plans to Place Salmon in Polluted Lake Ontario," *New York Times*, 15 September 1976, p. 1.

22. "Pollution Hurts Big World Rivers," *New York Times*, 21 April 1976, p. 27.

23. W. E. Farrell, "Uneasy Duluth Filtering Water," *New York Times*, 20 February 1976, p. 31.

24. "Glen Cove, L.I., Lawns Go Thirsty after Polluted Wells Are Closed," *New York Times*, 9 July 1977, p. C21; "Maine Town Is Warned on Water," *New York Times*, 14 January 1978, p. 8.

25. W. Pelon, B. F. Whitman, and T. W. Beasly, quoted in "River Water Induces Bacterial Mutation," *Science News* 111 (1977): 393.

26. J. W. Drake, "Environmental Mutagenic Hazards," *Science* 187 (1975): 503–14.

27. T. Page, R. H. Harris, and S. S. Epstein, "Drinking Water and Cancer Mortality in Louisiana," *Science* 193 (1976): 55–57; N. Wade, "Drinking Water: Health Hazards Still Not Resolved," *Science* 196 (1977): 1421–22.

28. L. Forrestal, "Deep Mystery," *Environment* 17 (1975): 25–32.

29. J. Crossland, "The Wastes Endure," *Environment* 19 (1977): 6–13.

30. Ibid.

31. K. A. Shuster, "Case Study of the Sayville Solid Waste Disposal Site in Islyp (Long Island), New York; Leachat Damage Assessment," U.S. Environmental Protection Agency Report #SW–509.

32. Drake, "Environmental Mutagenic Hazards"; Page et al., "Drinking Water and Cancer Mortality in Louisiana"; Wade, "Drinking Water."

33. Drake, "Environmental Mutagenic Hazards."

34. M. W. Browne, "Hazardous Chemicals Are Vital to Nation's Industry," *New York Times*, 28 February 1978, p. 18.

35. D. Zwick and M. Benstock, "Water Wasteland," report of the Ralph Nader study group on water pollution, 1972.

36. R. Raloff, "Water Pollution: Appearances Can Be Deceiving," *Science News* 112 (1977): 428–31.

37. J. Stein, "Fumbled Help at the Well," *Environment* 19 (no. 5, 1977): 14–17; ib., "Water for the Wealthy," *Environment* 19 (no. 4, 1977): 6–14.

38. "Ocean Disposal Site for Nuclear Waste Suggested," *New York Times*, 6 September 1977, p. 17.

39. R. A. Brown and H. L. Huffman, Jr., "Hydrocarbons in Open Ocean Waters," *Science* 191 (1976): 847–49.

40. D. Burnham, "Radioactive Material Found in Oceans," *New York Times*, 31 May 1976, p. 13.

41. C. Kaiser, "U.S. Says Army Sea Dump Perils New York Waters," *New York Times*, 31 July 1976, p. 1; "Study Finds High Metal Pollution in New York Waters Bare of Fish," *New York Times*, 26 September 1977, p. 35; "Study Says Fish Kill Off Jersey Signifies 'Major Damage' to Sea," *New York Times*, 31 August 1976, p. A9; R. R. Silver, "Nassau Beaches Closed as Sewage Incursion Widens," *New York Times*, 23 June 1976, p. 1.

42. R. Severo, "PCB Cleanup Cost Put at $20 Million," *New York Times*, 26 April 1976, p. 1.

43. Ibid.

44. "PCB Found in Fish on Market along East Coast," *New York Times*, 13 August 1976, p. A9.

45. B. A. Franklin, "Allied Chemical Given a Fine of $13 Million in Kepone Pollution," *New York Times*, 6 October 1976, p. 1.

46. F. S. Sterrett and C. A. Boss, "Careless Kepone," *Environment* 19 (1977): 30–37, and references to EPA reports cited therein.

47. "PCB Found in Fish on Market along East Coast."

48. H. M. Schmeck, Jr., "U.S. Program on Pesticide Safety Found in 'Chaos' in Senate Report," *New York Times*, 3 January 1977, p. 1.

49. L. J. Carter, "Chemical Plants Leave Unexpected Legacy for Two Virginia Rivers," *Science* 198 (1977): 1015–20.

50. B. J. Finlayson and J. N. Pitts, Jr., "Photochemistry of the Polluted Troposphere," *Science* 192 (1976): 111–19.

51. F. H. Pough, "Acid Precipitation and Embryonic Mortality of Spotted Sala-manders, *Ambystoma maculatum*," *Science* 192 (1976): 68–70.

52. C. R. Thompson, "Effects of Air Pollutants on Lemons and Navel Oranges," *California Agriculture* 22 (1968): 2–3. For additional references, see N. T. Keen and O. C. Taylor, "Ozone Injury in Soybeans," *Plant Physiology* 55 (1975): 731–33; R. F. Brewer, F. H. Sutherland, and F. B. Guillemet, "Effects of Various Fluoride Sources on Citrus Growth and Fruit Production," *Environmental Science and Technology* 3 (1969): 378–81; T. A. Mansfield, ed., *Effects of Air Pollutants on Plants* (Cambridge: Cambridge University Press, 1976).

53. "Pollution Control Shifts to Industry," *San Francisco Sunday Examiner Chronicle*, 25 November 1976, p. 6. The reference is to a quotation of Prof. James Pitts, Jr., director of the Air Pollution Research Center at the University of California at Riverside.

54. D. Pimentel, E. C. Terhune, R. Dyson-Hudson, S. Rochereau, R. Samis, E. A. Smith, D. Denman, D. Reifschneider, and M. Shepard, "Land Degradation: Effects on Food and Energy Resources," *Science* 194 (1976): 149–55.

55. W. S. Cleveland, B. Kleiner, J. L. McRae, and J. L. Warner, "Photochemical Air Pollution: Transport from New York City Area Into Connecticut and Massachusetts," *Science* 191 (1976): 179–82.

56. Pough, "Acid Precipitation and Embryon Mortality . . ."

57. P. R. Ehrlich and A. H. Ehrlich, *Population, Resources, Environment* (San Francisco: W. H. Freeman, 1970).

58. K. C. Sanderson, "Monitoring the Effects of Air Pollutants on Horticultural Crops," *Horticultural Science* 10 (1975): 489–504.

59. Damon and Kunen, "Global Cooling!"

60. V. Siegenthaler and H. Oeschger, "Predicting Future Atmospheric Carbon Dioxide Levels," *Science* 199 (1978): 388–95.

61. W. C. Wang, Y. L. Yung, A. A. Lacis, T. Mo., and J. E. Hansen, "Greenhouse Effects Due to Man-made Perturbations of Trace Gases," *Science* 194 (1976): 685–90.

62. G. M. Woodwell, "The Carbon Dioxide Question," *Scientific American* 238 (1978): 34–43.

63. Wang et al., "Greenhouse Effects Due to Man-Made Perturbations of Trace Gases."

64. Siegenthaler and Oeschger, "Predicting Future Atmospheric Carbon Dioxide Levels."

65. W. L. Gates, "Modeling the Ice-Age Climate," *Science* 191 (1976): 1138–44; CLIMAP Project Members, "The Surface of the Ice Age Earth," *Science* 191 (1976): 1131–36.

66. R. Bryson, "Climatic Modification by Air Pollution. II. The Sahelian Effect," IES Report #9, Center for Climatic Research, Institute for Environmental Studies, University of Wisconsin at Madison, 1973.

67. L. J. Carter, "Asbestos: Trouble in the Air from Maryland Rock Quarry," *Science* 197 (1977): 237–40.
68. A. N. Rohl, A. M. Langer, and I. J. Selikoff, "Environmental Asbestos Pollution Related to Use of Quarried Serpentine Rock," *Science* 196 (1977): 1319–1322.
69. H. O. Carter, "A Hungry World: The Challenge to Agriculture," summary report of the Division of Agricultural Sciences, University of California at Davis, 1974.
70. Ibid.
71. P. R. Crosson, "Institutional Obstacles to Expansion of World Food Production," *Science* 188 (1975): 519–24.
72. Pimentel et al., "Land Degradation"; E. P. Eckholm, *Losing Ground* (New York: W. W. Norton, 1976).
73. G. Hill, "United Nations Study Says Spreading Deserts Are Caused by Man's Misuse of the Land," *New York Times*, 25 February 1976, p. 17.
74. Pimentel et al., "Land Degradation."
75. Ibid.
76. Ibid.
77. E. P. Eckholm, "Salting the Earth," *Environment* 17 (1975): 9–15.
78. Ibid.
79. J. Kandell, "Brazil's Agriculture Expands Fast, But Mostly for Benefit of Well-to-do," *New York Times*, 16 August 1976, p. 2; "In a Hungry World, Brazil Emerges as Agricultural Giant," *New York Times*, 14 April 1977, p. 2.
80. G. Hill, "Environment Lag Is Found for '75," *New York Times*, 19 January 1976, p. 13; E. W. Kenworthy, "Industry Assails Water Standards," *New York Times*, 21 January 1976, p. 36; G. Hill, E.P.A. Backs Clean-Air 'Trade-off,' Allowing New Industrial Pollution," *New York Times*, 11 November 1976, p. 1; J. F. Sullivan, "Clean-air Rules Easing for South Jersey," *New York Times*, 12 February 1976, p. 1; W. Lissner, "Con Edison Gains Support for the Use of Dirtier Fuel," *New York Times*, 21 February 1976, p. 28; "Clean-water Rules Eased by Conferees," *New York Times*, 12 November 1976, p. 1.
81. Ibid.

Chapter 7
SECONDARY
NATURAL RESOURCES

The roots of industrial civilization lie in the basic primary natural resources whose future we have just evaluated. But these primary resources have significance to industrialism in part because they are the source of the secondary and eventually tertiary resources on which industrial civilization immediately depends. Secondary resources include animals, vegetable resources such as forest and agricultural products, and mineral resources like metal ores, gravel, and cement. One of these secondary resources—agricultural products—represents the output of the all-important food cycle, and lies at the heart of the transition envisioned in the second half of this book. Hence we shall defer discussion of agriculture to future chapters. In the present chapter we will examine the remaining secondary resources, and especially metal ores, with two goals. First, we shall consider the role of each secondary resource in the industrial process. Second, we shall attempt to project the future of the major secondary resources. In the process we shall see that industrial civilization faces a materials shortage whose long-term implications easily match those of the energy shortage.

Animal Resources

From the dawn of human history, fish, game, and wildlife have played a pivotal role in human existence. But the human species has placed unremitting pressure on its natural food supplies; as we have noted, the extinction of wildlife species by ancient human hunters is believed to have forced the growing human population to adopt the agricultural way of life. In the 20th century, humanity's natural capacity to affect its environment has been magnified manyfold by fossil fuels, and the extinction of wildlife has accelerated accordingly. Species after species of animal and plant has been extinguished by human activity, and dozens more are in imminent danger of extinction.[1]

In view of the rich diversity of life that has evolved on planet earth, extinction of a few dozen species may seem a minor dent in nature's armor, and irrelevant to the future of industrial civilization. After all, the extinction of the passenger pigeon earlier in this century, while attended by considerable nostalgia, caused not even a ripple in booming industrial economies. But here the fault lies not in logic, but paradigm. Nothing in nature is unrelated. Consider the case of pesticides, which we will see in future chapters are integral components of modern industrial agriculture as currently practiced. When pesticides are used, the earth is poisoned unselectively, killing not only "harmful" organisms but beneficial ones as well. For example, pesticides are steadily depleting the nation's honeybee population. In the decade from 1964 to 1974 the U.S. honeybee population declined by 20%, and mortality is rising. Already growers in California are forced to import bees from as far away as Wisconsin.[2] Fully one-third of the average American diet depends directly or indirectly on honeybee-pollinated crops; unless the destruction stops. U.S. agriculture will suffer significant losses in the next decade.[3]

Pesticides do not stop with honeybees: they enter natural food chains and are eventually incorporated into human tissue, especially fat. When this fat is later called on as a source of energy, the stored pesticides are liberated, flooding our bodies with poisons. The potent pesticide DBCP, produced and distributed by the Dow Chemical Company and used widely in California and Hawaii, is now suspected to cause human sterility and cancer.[4]

The effects of industrialism are not confined to a few species. The world's fisheries have been decimated in the 20th century by floating fish factories—commercial fishing ships that are equipped to catch, clean, and can their product in quantity, without regard to consequence. The Pacific anchovy industry is moribund; whales and porpoises are endangered; the San Francisco crab fishery is no longer commercially viable; and even the once-rich salmon fishery of the Pacific Northwest is being drained of its silvery treasure (Figure 7.1).[5] Vast tracts of the Atlantic Ocean that once served as important sources of seafood are now devoid of life as a consequence of unremitting pollution (Chapter 6). If present trends continue, many of the animal and plant species on which humanity now depends may be extinct or endangered by the turn of the century.

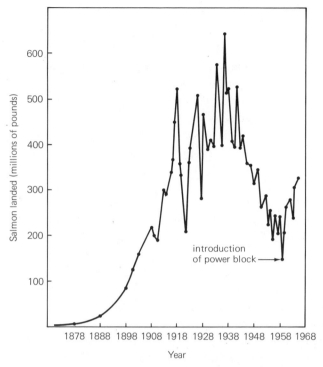

7.1 Alaskan salmon catch over the past century. Note the approximate bell shape of the curve, and the establishment of a new depletion curve with the introduction of the power block for hauling nets aboard. (data from C. H. Lyles, *Fishery Statistics of the United States, 1966* [Washington, D.C.: U.S. Government Printing Office, 1966], p. 517)

The Forest Resource
Use of the Forests

One of the most important materials to industrialism is the wood derived from forests. It is estimated that forests cover one-third of the earth's land surface, and produce nearly two-thirds of the oxygen we breathe.[6] In addition to their recreational and esthetic value, forests furnish raw material for the timber industry. Worldwide, timber is one of the largest industries; indeed, with the recent exploitation of the Amazon Basin,[7] there is no longer an uncut, virgin forest on the planet. In the U.S., timber is the fifth-largest industry.[8] In 1976 this industry harvested and processed more than 250 million tons of raw wood, more than one ton per year for every man, woman, and child in the U.S.[9] About one-third of this yearly harvest is converted to fiber-based products, such as pulp, paper, and cardboard, the remaining two-thirds is used for construction.[10]

From an energetic viewpoint, wood is an especially valued resource, because it requires but 2%–10% of the total energy input of other structural materials like steel and brick.[11] Moreover, it is technically possible to convert wood and other vegetable products into chemicals, plastics, rubbers, and even gasoline.[12] Such possibilities have led some authors to envision a coming Age of Wood.[13] It is theoretically possible for forests to fuel industrialism, for like fossil fuels trees contain stored solar energy. But unlike fossil fuels, trees represent a relatively low-grade energy form. One ton of wood contains about one-tenth the energy of a ton of high-grade coal. Given current technology, 150 million tons of wood—60% of the annual U.S. production—would be required to simply meet current U.S. demand for plastics and other polymers.[14] The production of such polymers in turn comprises but a few percent of the current U.S. energy budget.

Even if trees contained enough energy to power industrialism, it would not be possible on geographical grounds. The most productive forests in the U.S. are located in the Northwest and Southeast, far from the nation's industrial and population centers. The large-scale movement of forest products from production sites to consumption centers requires a relatively cheap and efficient transportation system, such as railroads. But the U.S. railroad system is in a state of deterioration,[15] having been replaced in the last 40 years by the highway system. The mismatch of internal production and consumption has made it economic for the U.S. to import wood, largely from nearby Canada.[16] From 1914 to 1976, wood imports increased from 0% to 14% of annual consumption, and the U.S. Forest Service projects a continuation of the trend.[17]

Depletion of the Forests

As long as fossil fuels are available in quantity to the timber industry, wood may be expected to continue to play an important role as a secondary natural resource. But in common with most natural resources, the forest resource has been depleted at an accelerating rate by industrialism. In Europe, the forest cover was reduced from 90% to 20% in the millennium from A.D. 900 to 1900.[18] In the U.S., forested area declined from 1600 to 1900 by nearly half, from two-thirds to less than one-third of the total land area, with a slight recovery after 1900 as former agricultural areas were abandoned.[19] Most significant, the tropical rain forests of the world, which account for about one-third of the planet's primary production of biomass,[20] are being reduced at an estimated annual rate of 0.5%–1.5%[21] as virgin rain forests are cleared for highways and agriculture.[22] At 1% per year tropical rain forests would be reduced to 75% of their present extent by 2007.

One consequence of the destruction of the world's forests that has been appreciated only lately is the increase of atmospheric carbon dioxide. As we saw in Chapter 6 (Figure 6.4), CO_2 in the air fluctuates seasonally, falling in the summertime as trees convert CO_2 and sunlight into biomass and rising again in the winter. When trees are eliminated, they no longer fix CO_2, which accumulates accordingly. Moreover, the carbon contained in wood is released to the atmosphere, further increasing CO_2 levels. It is now recognized that the CO_2 released from the destruction of forests is at least equal to that released from the combustion of fossil fuels,[23] with corresponding implications for the climate of the immediate future (Chapter 6).

The Future of Forests

Already the production of wood is approaching the saturation point in the sense that supply is having difficulty keeping up with demand. Experts predict that mounting pressures on the forest resource will reduce commercial timberland in the U.S. from 500 million acres today to 485 million by the year 2000 and 455 million by 2020[24] (Figure 7.2). Over the same time span the demand for timber products is projected by the U.S. Forest Services to increase exponentially, to more than twice the current demand (Figure 7.2). U.S. forests are theoretically capable of doubling in productivity over the next 50 years, but only with an immediate and massive commitment to intensive forest management.[25] Producing more wood from less land will require irrigation, drainage, and the heavy use of nitrogen fertilizers, all of which depend upon fossil fuels. It seems unlikely that such programs will be mounted in the face of declining fossil-fuel availability. Already the nonindustrial world has exhausted the firewood that has in the past supplied most of its energy.[26] In the industrial nations, the price of lumber appears to have reached the steeply rising portion of the exponential curve; in the summer of 1977 alone U.S. lumber prices increased at a rate of 168% per year.[27] The future promises more of the same. In the coming decades, timber—like all other natural resources—appears certain to increase in price and decrease in availability.

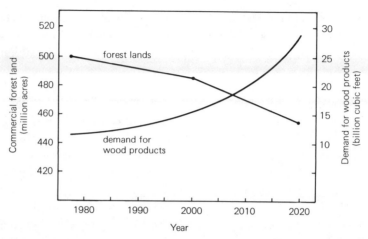

7.2 Projected acreage of U.S. commercial timberland, together with the projected U.S. demand for forest products. (acreage data from S. H. Spurr and H. J. Vaux, "Timber: Biological and Economic Potential," *Science* 191 [1976]: 752–56; demand projection from U.S. Forest Service data)

Mineral Resources

Long before industrial civilization destroys wildlife and the forests, it will confront its most severe materials shortage, occasioned by the depletion of mineral resources. These secondary natural resources include the ores from which metal is smelted, as well as nonmetal minerals such as sand, gravel, cement, gypsum, and asbestos. Industrial societies are by far the heaviest users of such resources. The U.S. alone is responsible for one-fourth of the world's annual consumption of minerals.[28] In 1974 this consumption rate amounted to more than 10 tons of mineral resources for every U.S. citizen, including 1,340 pounds of metals and 18,900 pounds of nonmetallic minerals.[29] At this rate each U.S. citizen consumes approximately 700 tons of mineral resources in a lifetime, including nearly 50 tons of metals. If we add fossil fuels[30] and wood[31] to this inventory, the individual lifetime consumption for each American more than doubles, to an astonishing 1,400 tons—a figure that still excludes food and water. The Industrial Age is truly the Age of Consumption.

At first glance it seems inconceivable that each of us could consume such a remarkable quantity of natural resources. And yet we need only examine our environment and our lifestyle to see where these resources have gone. Our cars, trucks, highways, cities, stadiums, and military machines are all made largely from mineral resources; and built, cooled, warmed, and fueled with fossil fuels. Moreover, once these edifices are erected, they must be maintained by the use of natural resources, at a staggering materials cost whose implications are seldom appreciated. Weather would destroy most of the U.S. highway system within a decade were it not maintained by the continuous application of asphalt, made largely from oil. Nickel, cobalt, and other metals are used to make high-temperature alloys that are strong but have short service lives. The steel from which the very foundation of our civilization is fashioned deteriorates rapidly and must be replaced constantly, at a current estimated annual cost of $20 billion.[32] In New York City alone, bridges built in some cases of steel cast before the turn of the century are crumbling and will require a third of a billion dollars to repair—a sum far beyond the budgetary capacity of the city.[33]

These brief examples illustrate a critical generality; industrial civilization is not the eternal physical edifice that we might imagine. Rather, it depends, like the human body, on a constant turnover of materials. Moreover, and also like the body, as industrial societies grow older they must divert an increasing proportion of their resources to maintenance. Even a no-growth economy therefore requires a steady and substantial stream of mineral resources. As we see next, the stream has begun to ebb.

The Depletion of Metals

The depletion of mineral resources is most significant and best documented in the case of metals. There is no disagreement that a continued supply of metals is essential to industrialism, and data on production histories and remaining reserves are relatively complete and accurate. Thus the future of metals can be evaluated with some confidence.

The Exponential Model

Like everything else in industrial societies the use of metals has increased exponentially in this century. In their book *Limits to Growth*, Meadows et al. estimated known global reserves of various metals and then assumed that the use of metals would continue to increase exponentially.[34] By this highly simplified model, industrial civilization will exhaust several key metals before the year 2000 (Table 7.1).

Table 7.1. FUTURE OF THE WORLD'S
METALS ASSUMING EXPONENTIAL
GROWTH FROM PRESENT
USE LEVELS [a]

Metal	Exhaustion date (exponential model)
aluminum	2003
chromite	2067
cobalt	2032
copper	1993
gold	1981
iron	2065
lead	1993
manganese	2018
mercury	1985
molybdenum	2006
nickel	2025
platinum	2019
silver	1985
tin	1987
tungsten	2000
zinc	1990

[a]Modified from D. H. Meadows et al., *The Limits to Growth* (New York: A Potomac Associates book published by Universe Books, 1972)

The Bell-Shaped Depletion Model

While *Limits to Growth* usefully focused attention on the impending metals crisis, we have seen that exponential growth is not a valid model for resource depletion. As we saw in the case of Easter eggs and fossil fuels (Chapter 3), the production of a natural resource is expected to follow a bell-shaped curve in which the availability of the resource grows exponentially, reaches a peak, and then declines. Such a depletion dynamic ensures that we will never fully exhaust any resource. Long before that time the declining concentration of the resource will make it uneconomic to recover. In this case the most relevant point on the curve is the time of peak production (the peak time), for that is when true and irreversible depletion begins in earnest.

The more realistic bell-shaped depletion model has been applied to metals by Arndt and Roper.[35] These studies represent an important step beyond Hubbert's similar earlier analysis of fossil fuels (Chapter 3), because the depletion of many metals is more advanced than fossil fuels. For such metals the depletion curve is therefore already established, and we can consequently predict the future availability of these metals with unprecedented accuracy. Certain of these highly depleted metals, such as gold, correspond as expected to the bell-shaped depletion model (Figure 7.3a). But other highly depleted metals, such as mercury, show a surprising deviation from the bell-shaped model (Figure 7.3b). In these cases the decline in the production curve is stretched out in time. This asymmetry implies that the metal has lasted somewhat longer than would have been predicted from the bell-shaped model.

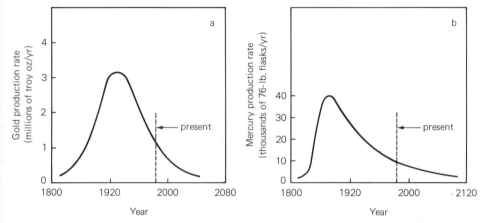

7.3   Production histories of gold (*a*) and mercury (*b*) in the United States. The depletion curve for gold (*a*) corresponds to the bell-shaped model of resource depletion, while the curve for mercury (*b*) is stretched out in time, corresponding to the asymmetric model. (from L. D. Roper, *Where Have All the Metals Gone?* [Blacksburg, Va.: University Publications, 1976], Figures 2 and 4)

The reasons for the asymmetry in the depletion curve are unknown. It is possible that the asymmetry is simply a sampling artifact, explained by the more comprehensive collection of production data since W.W. I. But it is also possible that the exponential growth of technology has enabled industry to wring the earth more thoroughly of its metal later in the production cycle, partially compensating the production decline. It is even possible that the economics of depletion stimulate extraordinary recovery efforts as depletion approaches; here, perhaps, is a central problem for the New Economists to solve. Finally, the asymmetry may result simply from increase in the discovery rate, which tends to lead actual production by 10–20 years. Regardless of the cause of the asymmetry in the depletion curve, an important effect is that the time of peak production becomes an ambiguous parameter of depletion. The time of peak production for the bell-shaped curve corresponds to the date at which exactly 50% of the resource is exhausted (the half date); but for the asymmetric curve, peak production can precede the half date by decades. The greater the asymmetry, the greater the difference between the peak time and the half date, which is why the peak date is ambiguous as a depletion parameter. Accordingly, Arndt and Roper[36] employed more accurate measures of depletion, derived from the "cumulative depletion" curve. Such a curve shows what percentage of the resource is exhausted by any given date (Figure 7.4). Relevant points on the cumulative depletion curve include the date at which half the resource is gone (the half date); the percentage gone at some recent reference date, say, 1975; and perhaps most telling for our purposes, the percentage of the resource that will be exhausted by some futue date, say, the year 2000.

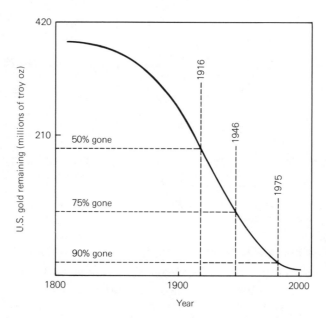

7.4 The cumulative depletion curve for gold in the United States. Such a curve shows how much of a given resource is exhausted by a given date. In this case, 90% of all U.S. gold had been mined by 1975. (from L. D. Roper, *Where Have All the Metals Gone?* [Blacksburg, Va.: University Publications, 1976], Figure 2)

These depletion parameters are summarized in Table 7.2 for several highly depleted U.S. metals. These metals include chromite, cadmium, manganese, and zinc, all essential to the manufacture of steel; mercury, silver, and the platinum group, all essential to the chemical and electrical industries; and lead and gold. Several other highly depleted metals are excluded from Table 7.2 because Arndt and Roper could not find reliable production data. These include beryllium, used in the nuclear and aircraft industries; niobium-tantalum, used in steel; and tin, essential to many modern alloys.

Table 7.2. DEPLETION PARAMETERS FOR HIGHLY DEPLETED U.S. METALS[a]

Metal	Peak date	Half date	% Gone in 1975	% Gone in 2000
cadmium	1957	1972	54%	75%
chromite	1955	1955	100%	100%
gold	1916	1916	90%	96%
lead	1925	1958	60%	71%
manganese	1955	1955	90%	98%
mercury	1870	1916	75%	80%
platinum group	1923	1941	82%	90%
silver	1908	1938	70%	99%
zinc	1943	1968	54%	67%

[a] Parameters defined in the text; data from L. D. Roper, *Where Have All the Metals Gone?* (Blacksburg, Va.: University Publications, 1976).

The crucial point of Table 7.2 is that the U.S. is rapidly depleting the metals on which an industrial economy depends. By the year 2000, many of these highly depleted metals will without question be exhausted from U.S. soil in the sense that they will no longer be economically recoverable. Moreover, the story does not end here. A second group of U.S. metals is "moderately depleted." Production of these metals has peaked only recently, and hence it is not yet possible to determine whether the depletion curve will be bell-shaped like gold, or asymmetric like mercury. Calculations based on both curves, however, furnish a range within which the actual depletion parameters are very likely to lie (Table 7.3). These moderately depleted metals include iron, the major source of steel; aluminum, used increasingly as a strong but lightweight structural material; and nickel and vanadium, indispensable in steel making. By the year 2000, most of these moderately depleted metals will be highly depleted (Table 7.3). Indeed, of U.S. metals that are essential to industrialism, only four appear far from peaking: antimony, copper, magnesium, and molybdenum. Peak U.S. production of these four metals is predicted from the bell-shaped model to occur between 2000 and 2020.[37]

**Table 7.3. DEPLETION PARAMETERS FOR MODERATELY
DEPLETED METALS IN THE U.S.[a]**

Metal	Range for peak date	Range for half date	Range for % gone by 1975	Range for % gone by 2000
aluminum (bauxite)	1968–1970	1966–1986	42%–66%	61%–92%
iron ore	1962–1974	1962–2021	29%–62%	40%–81%
nickel	1962–1987	1972–2010	17%–58%	39%–84%
selenium	1962–1964	1962–1974	51%–79%	74%–95%
tellurium	1964–1965	1964–1976	48%–75%	73%–93%
titanium (llemite, FeTiO$_2$)	1966–1967	1966–1979	44%–75%	74%–98%
titanium (Rutile, TiO$_2$)	1958–1959	1958–1967	61%–90%	82%–98%
tungsten	1962–1961	1962–1981	44%–72%	67%–94%
vanadium	1968–1972	1968–1989	33%–69%	61%–94%

[a] Each parameter is expressed as a range, the two extremes of which were calculated using the bell-shaped (symmetric) depletion model and the asymmetric model. Data from L. D. Roper, *Where Have All the Metals Gone?* (Blacksburg, Va.: University Publications, 1976).

Cornucopia Revisited

The message of Tables 7.2 and 7.3 would seem clear and unequivocal; we are running out of the metals that form the backbone of our industrial civilization. But there is an opposing view, a view that sees our earth as a boundless Cornucopian horn of plenty that can never be depleted of its treasures. Cornucopians believe metals are effectively unlimited in supply. The position is aptly summarized by Brooks and Andrews as follows:

> In short, almost every bit of evidence we have indicates the existence of vast quantities of mineral resources that could be mined and, further, that as either their price goes up or . . . technology of production improves . . . the volume of mineable material increases significantly—not by a factor of 5 or 10 but by a factor of 100 or 1000.[38]

Cornucopians point out that there are sufficient metal resources contained in the entire planet to last for literally millions of years at today's consumption levels. Thus, if the whole earth could be processed for its iron, it would yield an estimated 1.8×10^{15} tons, enough to last for 4.5 million years at 1968 consumption levels.[39] Cornucopians also note that the fraction of the total resource that is recoverable at existing prices and technology, termed the reserve, tends always to grow. For as a rich reserve is depleted, the price of the resource increases, making it profitable to mine increasingly leaner sources. Cornucopians are persuaded that this historic progression need never end, and hence that forecasts based upon static reserves (Tables 7.1 through 7.3) are meaningless.

The prospect of indefinite material abundance is alluring to many, and is an article of faith to such prominent futurists as Herman Kahn and Daniel Bell. Moreover, if we accept the Cornucopian position, our view of the future is radically altered. The challenge then becomes to manage indefinite economic growth, rather than to prepare for imminent economic decline. It is therefore crucial to look beneath the surface of this Cornucopian view.

The Economics of Resource Recovery

The Cornucopian position requires that as a resource is depleted, it becomes more expensive. On this count there is no disagreement. Mercury, for example, is rapidly approaching exhaustion in the U.S., and its production history reveals what may be the typical relation between resource availability and price (Figure 7.5). Since 1900, when mercury price and production were in economic equilibrium, the mean real price trend of mercury increased exponentially, with a doubling time of 20 years (Figure 7.5). The year-by-year price shows a regular oscillation that is superimposed on the exponential rise. When prices increase, more metal is produced after a variable time lag, as required by the Cornucopian argument.

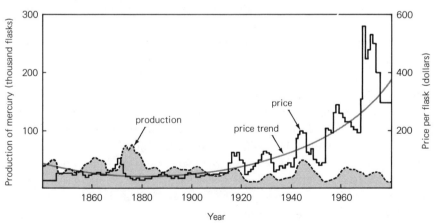

7.5 Production and price history of mercury in the United States. (from E. Cook, Texas A&M, "Limits to Exploitation of Nonrenewable Resources," *Science* 191 [1976]: 677–82, Figure 4)

The Technology
of Resource Recovery

Cornucopians can draw support also from the history of U.S. mining technology, which shows that estimated reserves have often increased over time. In 1934, for example, there were sufficient U.S. reserves of four major metals (copper, iron, lead, and zinc) to last for 15–40 years at the production levels which then prevailed. But by 1974, technological developments made it possible to mine lower grade ores. Consequently, the estimated U.S. reserves of the same four metals had doubled, to 24–87 years of production at the higher 1974 levels.[40]

The role that technology plays in increasing metal reserves is well illustrated in the production history of iron ore in the Lake Superior mining district (Figure 7.6). The production curve roughly corresponds to the now-familiar bell shape, with the exception of a slowdown during the Great Depression. In 1935, estimated reserves were sufficient for 20 years of additional production. A low-grade iron-bearing rock called taconite was not included in these reserves because it was considered unmineable at prewar prices and technology. But during W.W. II, technological advances born of necessity made taconite mineable. The iron contained in taconite was added to reserves, iron production was extended several years, and a new depletion curve for taconite was established (Figure 7.6).

The Law of Diminishing Returns

The Cornucopian view is thus supported by conventional economic wisdom; the increasing price of a resource stimulates technological development, which opens a greater fraction of the resource to exploitation. But while the Cornucopian view draws strength from one economic principle, it founders on another—the law of diminishing returns. In the case of fossil fuels we have seen that secondary and tertiary recovery techniques can extend production somewhat. But a recent comprehensive report by the U.S. Office of Technology Assessment, based on an investigation of every known U.S. oil reservoir, excluding those in Alaska, concluded that even assuming the most favorable market conditions, enhanced recovery techniques will increase the production rate by only 12%.[41] Likewise, the development of the taconite technology extended production in the Lake Superior district by years, but increased the total iron available from the district by only 10%. Similarly, in a typical copper mine, 90% of the copper metal present can already be extracted with ex-

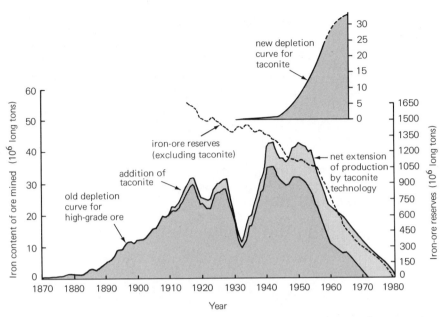

7.6 Depletion history of iron ore in the Lake Superior mining district, illustrating the overlapping stages of depletion of a mining region. The lower black curve shows the depletion of high-grade ore. The white overlapping curve shows the additional production made possible by the taconite technology. The curve inset at the upper right shows the new depletion curve for taconite. (from E. Cook, Texas A&M, "Limits to Exploitation of Nonrenewable Resources," *Science* 191 [1976]: 677–82, Figure 3)

isting technology. New techniques can therefore add at most 10% to the total copper produced.[42] In fact, nearly three-quarters of the copper metal contained in the most abundant type of North American copper ore is contained in high-grade (0.7%) deposits. Lower grade deposits comprise increasingly smaller, not larger, quantities of copper.[43] The same is true of virtually all deposits of fossil fuels and metal ores; the harder they are worked, the less they yield.

These examples illustrate what may seem obvious: once the water is squeezed from a sponge, squeezing harder produces little additional water. In the same way, lower grade deposits of metal represent increasingly smaller, not larger, total amounts of metal. Improved technology can without question make this additional metal available to industrialism, but the amounts involved are barely significant. Moreover, squeezing the earth harder for its increasingly diffuse metals requires exponentially increasing amounts of energy (Figure 7.7). Cornucopians readily concede that their position depends on finding a cheap, inexhaustible, and nonpolluting energy source.[44] In fact, no energy source is non-polluting, since all generate waste heat. Neither is future energy likely to be cheap, as we shall document in the next chapter. And as we have seen, "inexhaustible" energy does not exist, if only because all energy sources depend on exhaustible metals and other resources.

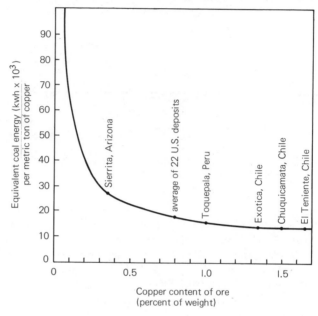

7.7 Equivalent coal energy required for mining different grades of copper sulfide ores. (from E. Cook, Texas A&M, "Limits to Exploitation of Nonrenewable Resources," *Science* 191 [1976]: 677–82, Figure 1)

The Cornucopian view is an illusion born of abundance. In the early stages of a resource production cycle, while the rate of new discoveries is still growing, the steady increase in estimated reserves that lies at the heart of the Cornucopian position is inevitable. But when the discovery rate peaks and begins to decline, as has occurred for many U.S. fuel and mineral resources, it is only a matter of time until reserves also begin to shrink. Already the estimated reserves of several U.S. resources have begun to decline, including oil, natural gas, silver, gold, platinum, and mercury. Within a decade the estimated U.S. reserves of other metals may be expected to follow suit.[45] Moreover, even Cornucopians agree that depletion implies relative increases in price. As I shall develop shortly, the soaring price of energy and resources may itself be sufficient to transform our industrial civilization.

The Age of Substitution

When a nation exhausts its resources, one of its options is substitution. Materials substitution is especially feasible in industries that use relatively small absolute quantities of resources, such as electronics.[46] Technology can also make available new materials that can be substituted for the depleted ones.[47] Thus materials derived from "renewable" resources can be substituted on a limited scale for exhausted fossil fuels and metals.[48] Technology can also help find a completely different approach to performing a desired function. The vacuum tube, for example, has been largely displaced by the transistor. Such functional substitutions can lead from dependence on a scarce material to the use of a more abundant one.

As mineral resources are exhausted in the coming decades, industrial societies will no doubt rely increasingly on substitution, both material and functional. Indeed, substitution is not new to industrialism; we have seen how the evolution of industrialism has been marked by the substitution of one energy source for another. But substitution has limits. To begin, certain crucial materials simply have no adequate substitute. It is impossible, for example, to conceive of industrialism on today's scale without a steady flow of high-grade structural steel; and there is no known substitute for chromium in producing such steel.[49] Second, substitutions are seldom as satisfactory as the materials they replace. Thus, while aluminum can be used in place of copper for many electronic applications, the electrical properties of aluminum are inferior and the energetic cost is several times higher.[50] Finally, many substitutions depend on technology; as we shall see in the next chapter, technology itself may be considered a resource that is subject to the same depletion dynamic as any other resource. In short, it is likely that the strategy of substitution can help alleviate the increasing scarcity of mineral resources, but the solution is bound to be inadequate and temporary.

Recycling

Mineral resources, like energy, follow the first law of thermodynamics. That is, they are never "used up" but rather converted from one form to another. As long as energy is available to power such conversions, many mineral resources, and especially certain metals, can be reused again and again. Such recycling can be less costly from an energetic viewpoint than obtaining metals directly from nature, since the energy expenditure necessary for the initial mining and smelting of ore need not be repeated.

Industrial nations already satisfy a significant fraction of their metal appetite through recycling. In the U.S., for example, recycling supplies about half the annual demand for antimony, one-third the demand for iron, lead, and nickel, and one-forth the demand for mercury, silver, gold, and platinum.[51] Municipal wastes provide an increasingly attractive source of raw materials for such recycling.[52]

As metals become increasingly scarce in the coming decades, recycling is expected to become more economic. But the second law of thermodynamics also applies to mineral resources; each time a metal is recycled, a portion is inevitably lost. Moreover, the technical, institutional, and financial barriers to recycling are formidable.[53] Future improvements are possible, but they cannot begin to compensate for the shortages that result from depletion. Reliable sources estimate that recycling can meet little more than 1% of the total U.S. demand for most mineral resources over the next decade.[54]

Foreign Dependence

When an industrial nation exhausts its own mineral resources, it can turn to other nations to meet its needs. Japan already imports all its fuel and most of its metal requirements, and the U.S. has been a net importer of metals since W.W. II.[55] In 1974 this nation depended on foreign suppliers for 98% of its manganese and cobalt, 91% of its chromite, 88% of its aluminum, 86% of its tin, and more than 50% of a dozen other critical metals (Figure 7.8). In several instances industrial nations are dependent for supplies of critical resources on unstable or politically hostile countries. Chromium, for example, which is required for making high-grade steel, is concentrated in South Africa. Phosphates, which are essential in fertilizers, are controlled by the small North African nation of Morocco. The dependence of industrial nations on imported mineral resources raises the specter of political blackmail and economic leverage of the kind that Arab nations now enjoy because of their oil.[56] Indeed, it would be surprising if nonindustrial nations did not employ such tools in their quest for economic equality.

But even if we ignore the potential dangers of dependency, industrial nations cannot count on importation of mineral resources for long. The scarcity of minerals is not localized to industrial nations; it is global in scale. World production of arsenic and silver has peaked, and at least seven other important metals are at or near world production peaks.

Mineral	Percentage imported						Major foreign sources
		0	25	50	75	100	
strontium	100						Mexico, UK, Spain
columbium	100						Brazil, Malaysia, Zaire
mica (sheet)	99						India, Brazil, Malagasy
cobalt	98						Zaire, Belgium, Luxembourg, Finland, Norway, Canada
manganese	98						Brazil, Gabon, South Africa, Zaire
titanium (rutile)	97						Australia, India
chromium	91						USSR, South Africa, Turkey, Philippines
tantalum	88						Australia, Canada, Zaire, Brazil
aluminum (ores and metal)	88						Jamaica, Australia, Surinam, Canada
asbestos	87						Canada, South Africa
platinum group metals	86						UK, USSR, South Africa
tin	86						Malaysia, Thailand, Bolivia
fluorine	86						Mexico, Spain, Italy
mercury	82						Canada, Algeria, Mexico, Spain
bismuth	81						Peru, Mexico, Japan, UK
nickel	73						Canada, Norway
gold	69						Canada, Switzerland, USSR
silver	68						Canada, Mexico, Peru, Honduras
selenium	63						Canada, Japan, Mexico
zinc	61						Canada, Mexico, Peru, Australia, Japan
tungsten	60						Canada, Bolivia, Peru, Thailand
potassium	58						Canada
cadmium	53						Mexico, Canada, Australia, Japan
antimony	46						South Africa, Mexico, P.R. China, Bolivia
tellurium	41						Peru, Canada
barium	40						Ireland, Peru, Mexico
vanadium	40						South Africa, Chile, USSR
gypsum	37						Canada, Mexico, Jamaica
petroleum (inc. nat. gas liq.)	35						Canada, Venezuela, Nigeria, Netherlands, Anti, Iran
iron	23						Canada, Australia
titanium (ilmenite)	23						Canada, Venezuela, Japan, Common Market (EFC)
lead	21						Canada, Peru, Australia, Mexico
copper	18						Canada, Peru, Chile, South Africa
pumice	8						Italy, Greece
salt	7						Canada, Mexico, Bahamas, Chile
magnesium (nonmetallic)	6						Greece, Ireland, Austria
cement	4						Canada, Bahamas, Norway, UK
natural gas	4						Canada

7.8 Importation of metals by the United States in 1974. (from R. C. Kirby and A. S. Prokopovitsh, "Technological Insurance Against Shortages in Minerals and Metals," *Science* 191 [1976]: 713–19, Figure 1)

These include cobalt, manganese, nickel, niobium, the platinum group, titanium, and tungsten.[57] Peak production of world minerals is centered around 1990, approximately 20 years behind U.S. production peaks. In other words, importation of mineral resources, like importation of fossil fuels, is at best a short-term solution to scarcity.

Metals and the Future

The demand for mineral resources is projected to triple by the year 2000,[58] corresponding to an annual growth rate of 5%. For select metals, the projected growth rate is even higher; thus the demand for bauxite, the major source of aluminum, is expected to grow by 9% per year, corresponding to an eightfold increase by 2000.[59] New mineral resources will certainly be discovered in the coming decades, but at a decreasing rate. The additions to reserves of many metals have been declining steadily since 1960 despite a steady increase in exploration.[60] The geological conditions that yield rich ore deposits are well known and the surface of the earth has been largely explored. Geologists tell us that there are not many new sources of minerals left to discover.

Once the earth's land masses are depleted of their high-grade ores, industrial societies may turn to the oceans, as they have already done in the case of fossil fuels. The oceans may indeed yield some metals: in particular, manganese nodules—small lumps of uncertain biological origin that contain not only manganese but also iron, nickel, copper, and cobalt—occur on the seabed in many parts of the world.[61] The economic viability of mining these nodules is not yet established, however, and only a few areas on earth have nodules with a sufficiently high metallic content to sustain a commercially viable mining effort.[62] Proposals to mine the moon for its resources and to capture metal-rich asteroids have not progressed beyond the realm of science fiction.

The weight of evidence thus indicates that within the foreseeable future the demand for several metals will outstrip the supply. Substitution, recycling, and importation can help to reduce the immediate impact of scarcity, but they only delay the inevitable. It seems certain that by the year 2000 the steady stream of secondary natural resources that has sustained industrialism in the past will have begun an irreversible decline, resulting in steadily escalating prices.

NOTES

1. G. T. Prance and T. S. Elias, eds., *Extinction Is Forever* (New York: New York Botanical Garden, 1976).

2. J. M. Winski, "If You Don't Have Enough Woes, Try Fretting about Bees," *Wall Street Journal*, 7 November 1974, p. 20.

3. H. M. Caine, "Pesticides and Pollination," *Environment* 19 (1977): 28–33.

4. "Dow Chemical Is Trying to Recall Supplies of a Suspected Pesticide," *New York Times*, 26 August 1977, p. A22.

5. C. H. Lyles, *Fishery Statistics of the United States, 1966* (Washington, D.C.: U.S. Government Printing Office, 1966), p. 517.

6. R. H. Whittaker and G. M. Woodwell, *Productivity of Forest Ecosystems* (Paris: UNESCO, 1971), pp. 169–175.

7. J. C. Jahoda and D. L. O'Wearn, "The Reluctant Amazon Basin," *Environment* 17 (1975): 16–30.

8. E. C. Jahn and S. B. Preston, "Timber: More Effective Utilization," *Science* 191 (1976): 757–61.

9. Ibid.

10. Ibid.

11. J. S. Bethel and G. F. Schreuder, "Forest Resources: An Overview," *Science* 191 (1976): 747–52.

12. I. S. Goldstein, "Potential for Converting Wood into Plastics," in *Materials: Renewable and Non-renewable Resources*, edited by P. H. Abelson and A. L. Hammond (Washington, D.C.: American Association for the Advancement of Science, 1976), pp. 179–84; K. V. Sarkanen, "Renewable Resources for the Production of Fuels and Chemicals," *Science* 191 (1976): 773–76.

13. E. Glesinger, *The Coming Age of Wood* (New York: Simon & Schuster, 1949).

14. Goldstein, "Potential for Converting Wood into Plastics."

15. E. Holsendorph, "Adams Say Neglected Roadbeds Are Key Factor in Rail Accidents," *New York Times*, 28 February 1978, p. 18.

16. Bethel and Schreuder, "Forest Resources."

17. Ibid.

18. H. C. Darby, cited in G. M. Woodwell, "The Carbon Dioxide Question," *Scientific American* 238 (1978): 34–43.

19. C. S. Wong, "Atmospheric Input of Carbon Dioxide from Burning Wood," *Science* 200 (1978): 197–200.

20. Woodwell, "The Carbon Dioxide Question."

21. Ibid.

22. Jahoda and O'Wearn, "The Reluctant Amazon Basin."

23. Wong, "Atmospheric Input of CO_2"; Woodwell, "The Carbon Dioxide Question."

24. S. H. Spurr and H. J. Vaux, "Timber: Biological and Economic Potential," *Science* 191 (1976): 752–56.

25. Ibid.

26. E. P. Eckholm, *Losing Ground* (New York: W. W. Norton, 1976).

27. "H.U.D. Seeks Lumber Study," *New York Times*, 28 August 1977, p. 27.

28. R. C. Kirby and A. S. Prokopovitsh, "Technological Insurance Against Shortages in Minerals and Metals," *Science* 191 (1976): 713–19.

29. Ibid.

30. S. V. Radcliffe, "World Changes and Chances: Some New Perspectives for Materials," *Science* 191 (1976): 700–7.

31. John and Preston, "Timber: More Effective Utilization."

32. Kirby and Prokopovitsh, "Technological Insurance Against Shortages in Minerals and Metals."

33. G. Lichtenstein, "New York Bridges Aren't Falling, But Some Are Crumbling," *New York Times*, 27 March 1978, p. B1.

34. D. H. Meadows, D. L. Meadows, J. Randers, and W. W. Behrens III, *The Limits to Growth* (New York: Signet, 1972).

35. R. A. Arndt and L. D. Roper, *Depletion of United States and World Mineral Resources* (Blacksburg, Va.: University Publications, 1976); L. D. Roper, *Where Have All the Metals Gone?* (Blacksburg, Va.: University Publications, 1976).

36. Ibid.

37. Ibid.

38. D. B. Brooks and P. W. Andrews, "Mineral Resources, Economic Growth and World Population," in *Materials: Renewable and Non-renewable Resources*, edited by P. H. Abelson and A. L. Hammond (Washington, D.C.: American Association for the Advancement of Science, 1976), p. 42.

39. H. E. Goeller and A. M. Weinberg, "The Age of Substitutability," in *Materials: Renewable and Non-renewable Resources*, edited by P. H. Abelson and A. L. Hammond (Washington, D.C.: American Association for the Advancement of Science, 1976).

40. Arndt and Roper, *Depletion of . . . Mineral Resources.*

41. S. Rattner, "Complex Fuel-Recovery Methods Are Found Unlikely to Yield Much," *New York Times*, 18 July 1977, p. 12.

42. E. Cook, "Limits to Exploitation of Nonrenewable Resources," *Science* 191 (1976): 677–82.

43. J. W. Whitney, "A Resource Analysis Based on Porphyry Copper Deposits and the Cumulative Copper Metal Curve Using Monte Carlo Simulation," *Economic Geology* 70 (1975): 527–37.

44. Goeller and Weinberg, "The Age of Substitutability."

45. P. W. Gluild, "Discovery of Natural Resources," *Science* 191 (1976): 709–13.

46. A. G. Chynoweth, "Electronics Materials: Functional Substitutions," *Science* 191 (1976): 725–32.

47. W. B. Hillig, "New Materials and Composites," *Science* 191 (1976): 733–38.

48. Bethel and Schreuder, "Forest Resources."

49. V. K. McElheny, "Technology: Ways to Overcome a Chromium Embargo," *New York Times*, 5 April 1978, p. 47.

50. Chynoweth, "Electronics Materials."

51. S. L. Blum, "Tapping Resources in Municipal Solid Waste," *Science* 191 (1976): 669–75.

52. J. G. Abert, H. Alter, and J. F. Bernheisel, "The Economics of Resource Recovery from Municipal Solid Waste," in *Materials: Renewable and Nonrenewable Resources*, edited by P. H. Abelson and A. L. Hammond (Washington, D.C.: American Association for the Advancement of Science, 1976), pp. 54–60.

53. Y. M. Ibrahim, "Roadblocks for Recycling: Taxes and Transport costs," *New York Times*, 14 December 1977, p. 63.

54. Blum, "Tapping Resources in Municipal Solid Waste."

55. Kirby and Prokopovitsh, "Technological Insurance Against Shortages."

56. E. R. Fried, "International Trade in Raw Materials. Myths and Realities," *Science* 191 (1976): 641–46.

57. Arndt and Roper, *Depletion of . . . Mineral Resources.*

58. B. Varon and K. Takeuchi, "Developing Countries and Non-fuel Minerals," *Foreign Affairs* 52 (1974): 497–510.

59. Ibid.

60. Guild, "Discovery of Natural Resources."

61. A. L. Hammond, "Manganese Nodules: A Mineral Resource on the Deep Seabed," in *Materials: Renewable and Non-renewable Resources*, edited by P. H. Abelson and A. L. Hammond (Washington, D.C.: American Association for the Advancement of Science, 1976), pp. 117–18.

62. A. L. Hammond, "Prospects for Deep Sea Mining of Manganese Nodules," in *Materials: Renewable and Non-renewable Resources*, edited by P. H. Abelson and A. L. Hammond (Washington, D.C.: American Association for the Advancement of Science, 1976), pp. 119–21.

Chapter 8
THE IMMEDIATE LIMITS TO GROWTH

The supply of energy and resources of course sets the ultimate limit on industrialism, but in practice it is not the most immediate limit. As noted in earlier chapters, the dynamics of resource depletion ensure that we shall never "run out" of energy and resources; long before that time other limits will assert themselves, individually and in concert, to transform industrial civilization. In this chapter we shall examine these "other limits," with three goals in mind. First, we will develop a unified conceptual framework within which these limits can be understood and their common features recognized. Second, we shall examine the limits themselves, individually at first and then collectively. Third, we shall see that these limits operate in unsuspected alliance to form a complex, interactive web that represents the most immediate limit to the continuation of industrialism as we have known it.

The Nature of Limited Processes

All processes, whether they are chemical reactions between molecules in our bodies, the growth of populations, or the birth and the death of star systems, are regulated by limits. Moreover, these apparently diverse processes are limited in a conceptually identical manner. The universal rules that govern limited processes have been formalized by a small but important branch of mathematical engineering known as *control theory*.

144

While we need not here pursue details, a brief look at some basic elements of control theory will help us to appreciate how the limits to industrialism operate.

Positive Feedback

Our earlier discussions of positive feedback furnish an introduction to control theory. As we have noted, positive feedback is distinguished by its property of self-reinforcement. That is, a cause yields an effect that in turn acts as the cause, in a continuous cycle that can produce runaway or exponential growth. Positive feedback has played a central role in the evolution of industrialism, as we have noted. Fossil fuels are employed to amplify human labor, which increases per capita agricultural productivity, freeing yet more labor for manufacturing, which results in more fossil-fuel use, and so on.[1]

Negative Feedback

Although positive feedback has played an important role in the growth of industrialism, we know that it cannot continue indefinitely. Eventually, all self-reinforcing processes become self-limiting, in a process known as negative feedback. Consider, for example, population size. Positive feedback can yield the rapid growth of natural populations, but the growth itself depletes food and other essential resources, causing populations eventually to level off and decline. In other words, positive feedback sustains growth, but the growth process itself initiates negative feedback that eventually arrests the growth (Figure 8.1). Ecologists have documented this process for many natural populations of animals and plants.[2]

Whenever negative feedback occurs, it takes a finite time to make itself felt. When negative feedback is used to regulate electronic circuits, for example, the feedback delay is measured in fractions of a second. In population growth, the negative feedback engendered by food depletion may take months or years to become manifest; and in cosmic phenomena such as the birth and death of stars, negative feedback may operate on a time scale of centuries or millennia. Regardless of the length of the feedback delay, however, the consequence is the same: oscillation of the limited process. To understand how this oscillation is produced, consider population growth. During the initial stage food is abundant and positive feedback is unchecked, much as industrialism has been unchecked for the past two centuries. During this growth phase the population has no trouble staying ahead of its limits. But once the population reaches a certain

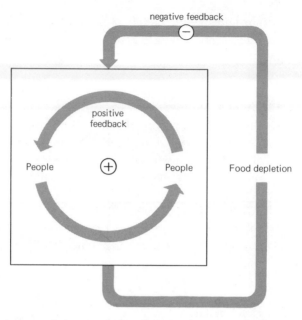

8.1 Feedbacks in population growth. Initially a population grows by positive feedback (inside box), but this growth must be checked eventually by negative feedback when food and other essential resources are depleted as a consequence of growth.

size the limits begin to catch up, manifest in the depletion of food. As a result, the population levels off and declines. But as it declines the limits are partially released, causing a new spurt of growth. The result of this see-saw dynamic is oscillation in the size of the population about some mean value, or "set point," that is determined ultimately by the environment's capacity to support the species.

The mathematics of such oscillation have been worked out in elaborate detail.[3] For our purposes, the point is simply that as positive feedback encounters the limits set by negative feedback, the result is oscillation. The first oscillation is usually the largest and subsequent peaks are progressively smaller, resulting in what engineers called "damped" oscillation. Soon we will see how these features of negative feedback may apply to the limitation of industrialism.

Environmental Limits and Negative Feedback

We have seen how industrialism grows by positive feedback; to see how it is constrained by negative feedback, we may make an analogy with the foregoing example of population. Whereas population growth depends on food, industrialism depends on energy and resources. And whereas population growth is limited by food depletion, industrialism is limited similarly by negative feedback as energy and resources become depleted.

But we have said that industrialism faces more immediate limits than energy and resource depletion. To see how these limits operate by negative feedback, recall that industrialism produces more than just manufactured goods: industrialism also generates such unwanted environmental by-products as air pollution. As documented in Chapter 6, air pollution reduces food production. When there is less food, there are eventually fewer people to work the levers of industrialism and to consume its many manufactured products. Thus while industrialism is self-reinforcing initially, it must eventually become self-limiting, by means of negative feedback (Figure 8.2).

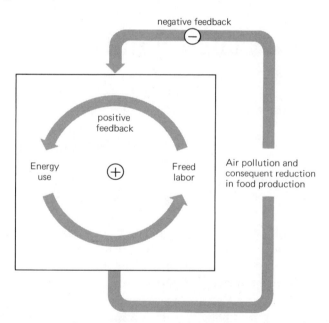

8.2 Feedbacks in the industrial process. Initially industrialism is sustained by positive feedback (inside box), in which energy use diverts labor from agriculture to manufacturing, which stimulates still greater energy use. This cycle is self-arresting, however, because its inevitable by-products counteract the process by negative feedback. In this example, air pollution reduces food production and thus limits the available labor.

In the above example, I have illustrated the effect of negative feedback using only a single by-product of industrialism, air pollution. In fact, the effects of air pollution on food production, although significant, are not yet nearly large enough to limit population effectively. But air pollution does not operate alone to limit industrialism; other environmental consequences of industrialism that we have discussed include climate modification by waste heat and carbon dioxide, water pollution, and land degradation. Each of these environmental limits also contributes to negative feedback regulation of industrialism. In reality, then, there is not one, nor even a few, but literally hundreds of sources of such negative feedback. And as we have noted, everything in the ecosystem is related to everything else. The limits that we are discussing therefore do not operate in isolation, but rather in a broad collaborative pattern. Our eventual goal in this chapter will be to grasp, as best the human mind can, the implications of this pattern. As a step toward this goal, let us examine in isolation several crucial categories of negative feedback limits that we have not yet considered.

The "Lead Time" Limit

We know that industrial civilization will, at its present pace, exhaust the energy and natural resources on which it depends within a few decades. We cannot predict the future with certainty, but we can be sure that major change will occur—either to a new energy source or to a new type of civilization. To prepare for these changes in advance requires lead time; but as time passes, the amount of lead time that is available to industrial societies of course diminishes.

The major source of lead time is the economic need to use existing capital facilities until they have returned their initial investment. A utility company that builds a nuclear power plant, for example, proceeds under the assumption that the plant will operate long enough to repay the initial capital outlay required to construct the plant, together with a sufficient profit to permit maintenance and perhaps expansion. In the absence of the expected return, the company would cease to function. Thus a commitment to build a power plant is really a commitment to build the plant and operate it for some time—typically 20–30 years. It is partly for this reason that past changes to a new energy technology have required a half century.[4]

In the case of nuclear power, objections to its widespread use in the U.S. have increased the amount of time that it takes to bring a plant into operation, from the technical minimum of 3–4 years to an average of 10 years or longer. As a result of such delays, construction costs increase, slowly at first but then more rapidly, in a typical exponential pattern.[5]

The cost overruns that have become so characteristic of large-scale technological enterprises have been caused in large part by exactly such delays. To formulate this aspect of the "lead time" problem in terms of the general concepts we have developed, we may conceive of the initial investment made by a power company as part of a positive-feedback process. Over time the investment yields returns, which can be reinvested to yield greater returns, and so on. Time delays cause increased costs, however, which have the effect of reducing the net return on an investment. Such increased costs take the form of extended interest payments on capital loans, extended payments on labor contracts, inflationary increases in the cost of labor and materials, etc. Time delays thus comprise a negative-feedback stabilization, or limit, on investment (Figure 8.3). This financial limit is making nuclear power increasingly unattractive to investors, with the consequence that 85% of the 170 reactors planned or under construction in the U.S. in the early 1970s have now been cancelled or deferred.[6]

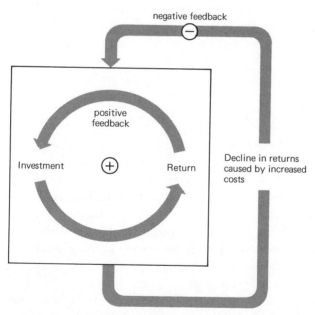

8.3 Feedbacks in economics. Investments made by power companies grow by positive feedback (inside box), in which investments yield returns which permit more investments. This growth process itself generates social resistance, which is translated into increased costs by construction delays. The increased costs in turn arrest investment by negative feedback.

The resistance to change that is imposed by time takes many forms. As we noted earlier, for example, solar energy is now economically competitive with electricity for heating homes in certain regions of the United States. Solar heating is most economic, however, when it is installed in new homes, but the turnover rate for homes in the U.S. is only 1% per year. That is, each year only 1% of the existing housing stock is replaced by new dwellings. Thus, even if every new home built in the U.S. between now and the year 2000 used solar heat, only one-fifth of U.S. homes would be heated by the sun in 2000.[7] Lovins has made a comparable calculation for nuclear power.[8] If the U.S. constructed a huge (1 gigawatt, or 1 billion-watt) nuclear reactor every day from the early 1970s to the year 2000, nuclear power would still meet only half of the projected energy needs of the U.S., given a continuation of historic growth patterns. In practice, such a gargantuan construction effort would require a social and economic mobilization comparable to that of World War II.

These brief examples illustrate that owing to the sheer magnitude of industrial civilization, it can change only slowly. Already the required lead time is approaching the theoretical minimum set by the need to find and then deploy new energy sources within a few decades. At the very least, the steady reduction of available lead time constitutes an increasingly potent negative-feedback limitation on industrialism.

Sociological Limits

Many authors have elaborated the view that industrialism is constrained not so much by the shortage of energy and resources as by ingrained institutional patterns that prevent effective reaction to the shortage.[9] To illustrate, consider the complexities of land-ownership patterns and their effect on the expanded use of coal as an energy source. Lands on which coal could profitably be mined are frequently broken into "checkerboard" patterns by land grants to railroads, ownership of coal rights by Indian tribes and private individuals, and leasing practices of the federal government (which is the principal owner of coal rights). As a result of these sociological patterns of land ownership and use, it can take 10 years from the initiation of a lease sale to the actual production of coal.[10] The high cost of such delay discourages the use of coal, and illustrates how sociological patterns of land ownership can impede changes to a new source of energy.

Local building codes that regulate construction in most of the U.S. can have a similar effect. While such codes are a convenient way for society to manage growth and establish construction standards, they can also discourage adaptive innovations such as nonconventional heating, nonflush toilets, water-recycling systems, etc. Again we see that institutional patterns that make sense in one context can also serve to retard beneficial change.

One of the most serious of such "sociological" limits to industrialism involves the system of transportation in the U.S. For a variety of socioeconomic reasons, U.S. society has progressed steadily away from energy-efficient modes of transportation such as railroads and waterways toward such energy-intensive modes as automobiles, trucks, and airplanes. The result of this trend has been the construction of millions of miles of highways in the U.S., to the relative neglect of the railroad system. As a result, the U.S. now finds itself critically dependent on automobile and truck transportation, especially in the newly industrialized South and Southwest. Oil and natural gas can be economically transported by truck, but the relatively bulky coal cannot. Thus institutional forces have helped create a transportation system well suited to the use of oil and gas, but inappropriate to coal. Modernization of the U.S. railroad system, which would be required if coal is to substitute for oil and natural gas on a large scale, would take several billions of dollars and several decades. Given the "lead time" limit, adequate modernization may not be possible even if it were economically feasible.

Political Limits

The political organization of industrial nations, and in particular the U.S., has proved a major institutional barrier to effective energy policies. One federal energy policy (domestic taxes) increases the flow of cash into petroleum production while another (price regulation) decreases it. One policy (taxes) encourages foreign oil production while another (import quotas) prevents the subsidized production from entering the U.S. market. One government agency (the Federal Energy Administration) orders 74 oil-fired plants to switch to the more abundant coal while another agency (Environmental Protection) succeeds in stopping all but one of the conversions because they will cause too much pollution.[11] A comprehensive study of 11 major federal energy policies over the last half century reveals a consistent pattern of "conflicting and counterproductive government policies."[12] Energy economists have described the situation as "organized waste."[13] The present administration's "Energy Program" reflects apparent awareness of the crisis that confronts industrialism, but a continued inability to cope effectively.[14]

The federal mismanagement of energy and resources is not caused by ill-will or incompetence. Rather, it is symptomatic of a deep and potentially intractable flaw that may prove the undoing of the American democratic ideal, namely, the sensitivity of government to the organized pressure of special-interest groups. Our political system selects and advances politicians that are maximally responsive to external pressures. Oil companies and other established industries control immense wealth, and can therefore mount irresistible pressures on politicians in the myopic pursuit of short-term economic gain. Moreover, even when socially useful legislation is passed, the wealthy oil concerns can challenge it in court; the U.S. legal system is also maximally responsive to money, since the most effective and proven lawyers command the highest fees. The wealth–power cycle[15] feeds on itself by positive feedback: wealth generates power which generates more wealth. Unless the cycle is interrupted it must eventually devour itself, like the proverbial Chinese dragon, but in the meantime it is a source of negative feedback that preserves a maladaptive status quo.

Institutional Homeostasis

Sociopolitical forces that tend to resist change permeate the fabric of all social institutions and societies. Some of these forces develop unintentionally, but others are deliberate, and for good reason. Without resistance to change, the stability and continuity that society provides its members would be impossible. Indeed, our bodies operate on the same principles to maintain essential physiological stability. When we are chilled, for example, a thermostat in our brain sends messages that cause our muscles to contract rhythmically. The heat produced by the resultant shivering serves to counteract the cold and maintain a constant internal temperature. Such physiological homeostasis, as it is termed, operates by negative feedback to maintain the constant internal conditions that are required by life processes. Similar resistance to change operates on the behavioral level, inducing us to prefer familiar diets, drive the same route to and from work each day, shop in the same markets, and so on. Resistance to change on the level of the individual is well documented by psychologists;[16] indeed, Freud himself was eventually convinced that a tendency to repeat past behavior patterns is stronger even than his "pleasure principle."[17]

Societies, like the body and the brain, have also necessarily evolved institutional devices that resist rapid change. Otherwise they would have ceased to exist. Moreover, such institutional homeostasis, like physiological homeostasis, is maintained by negative feedback. To illustrate, the U.S. legal system is based on precedent, that is, the outcome of past legal cases. Such a system, while open to slow change, is obviously and deliberately resistive to rapid change. It has built-in inertia. Likewise, patterns of land tenure and ownership, occupational seniority, unionization, moral and religious codes, the influence of "established" interests, the self-reinforcing interaction between power and wealth—all conspire to resist rapid change in favor of the status quo.

Throughout most of recorded history, external change has been relatively slow and institutional homeostasis has evolved to help provide a stable environment for social interchange and orderly cultural evolution. But for the past two centuries unchecked positive feedback has reigned, with the result that external change is now more rapid than at any time in human history. Society's usual internal mechanisms for change cannot keep pace. As a result, the resistance to change that is normally beneficial now serves to block rapid change when it is most needed. It is a fact of history that the speed with which a society can react to external threats is inversely related to the number of people who share decision-making power. By far the fastest reactions come from dictatorships. As noted by others,[18] the next few decades seem likely to test the capacity of a democratic form of government to respond adaptively to rapid change.

Economic Limits

Many of the limits we have been discussing eventually assert themselves through economics. Indeed, one of the most difficult immediate problems faced by industrial society is an economic one. Briefly, the problem is to find enough capital to sustain and expand the use of increasingly scarce fossil fuels while simultaneously developing and deploying new energy sources.

The High Price of Industrialism

When the oil industry was born a little over a century ago, huge quantities of oil could be obtained by merely pricking the surface of the earth. Now is is necessary to build multimillion-dollar oil platforms at sea and to construct billion-dollar pipelines that span continents. One consequence of energy depletion is escalation of the amount of energy, and therefore capital, that must be invested to deliver energy to world markets. Already it is estimated that one-fifth of the U.S. energy budget is devoted simply to bringing energy to market.[19] Over the next decade alone, it is estimated by one source that the world oil industry will require $2 trillion to meet projected demand,[20] more money than was spent by the entire world for military purposes in the preceding decade. The U.S. oil industry alone will require an estimated $300 billion,[21] while the electric utilities will need an additional $363 billion.[22] The U.S. nuclear industry, according to one estimate, could require $1 trillion through the year 2000.[23]

The capital requirements of the energy industry are unprecedented, and informed opinion is strongly divided over whether the necessary money can be generated. The Ford Foundation Energy Project, for example, concludes that the capital needs of the oil industry and electric utilities can be met through internally generated profits and borrowing from external capital markets.[24] This opinion is, however, openly based on orthodox economic theory, including the premise that "In a market where prices are free to equate supply and demand, there is no such thing as a capital shortage."[25] As developed in Chapter 5, this position fails to acknowledge thermodynamic realities now facing industrial nations. As we have seen, capital is not infinite in supply: rather, the money cycle is driven by the conversion of energy. When there is an energy and resource shortage, the resource cycle slows down, creating a capital shortage.

The burden of raising capital for energy development in the U.S. is expected to fall on private enterprise.[26] The rate at which new energy technologies can be developed and deployed by private enterprise is limited, however, by the amount of investment capital that can be generated or diverted from other uses. Even in times of economic growth, such investment capital seldom expands faster than a few percent per year. As documented in preceding chapters, the period of economic growth may soon belong to history, in which case capital growth will cease.

In the likely event that private enterprise is unable to raise sufficient capital to meet future energy needs, the support of the federal government will be the only recourse. The federal government has already underpinned the energy industry with such supportive legislation as the oil-depletion allowance and the Price–Anderson Act.* But raising a trillion

---

* This act limits the liability of utilities to $500 million in the event of a major nuclear accident. Without such protection, private industry was unwilling to invest in nuclear power.

extra dollars in two decades is quite another matter. Federal budgets typically operate so close to their limits that even in times of growth sufficient new capital for energy development could not be raised without major diversion from other sources, such as the military. And yet such diversion would itself alter industrial economies beyond recognition.

The Future Price of Energy

The escalating capital requirements of the energy industry will of course manifest themselves as an increase in the price of energy. While we can be certain that future energy will cost more, we cannot know in advance the precise shape of the curve. No in-depth studies on future energy prices have been performed,[27] and even if they were they would contain major imponderables. For example, much of the world's oil is controlled by a few Arab nations, banded together in economic alliance (OPEC). This alliance can defy conventional economic practice by unilaterally raising prices, as it demonstrated when it quadrupled oil prices in the early 1970s. Indeed, our own Federal Power Commission unexpectedly tripled the price of natural gas in July of 1976.*[28] In estimating the future price of energy, we can do little better than examine past trends and extrapolate them forward in time, supplemented by a survey of informed judgment.

What happens to the price of a valued commodity when it is depleted? While we cannot answer this question for oil, since it has never been depleted, we can seek guidance from other examples. Mercury, for instance, is essentially exhausted in the U.S. As we saw in the last chapter (Figure 7.5), the price of mercury has increased exponentially since 1900, with a doubling time of about 20 years. Superimposed on the exponential climb is a price fluctuation that roughly parallels the supply: when production goes up, more mercury is available, and so prices go down after a variable time lag, as expected from conventional price theory.

Assuming that the example of mercury reveals a general relation between the depletion of a resource and its price, we may expect the price curve for energy to oscillate in parallel with supply. That is, we may expect the world energy market to fluctuate between periods of glut and scarcity. This oscillation may reflect the characteristics of negative feedback discussed earlier in this chapter, operating in this case in the economic sphere. Whatever the cause, one of the consequences is periodic apparent abundance that can obscure a general trend toward scarcity and create a false sense of comfort that impedes deliberate corrective re-

---

*It later reduced the increase by 25%, admitting error in its earlier increase.

sponse. The recent world oil glut may provide an example of exactly this effect.

In the longer term, memory smooths the oscillations, and it is the overall or average price trend that is most relevant. In 1973, a conservative but informed estimate held that the average price of energy would double or triple by the year 2000,[29] corresponding to an annual price increase of 3%–4%. A 1976 report by the Energy Research and Development Administration of the U.S. government assumed an average annual increase in energy price of 10%, corresponding to an approximate quadrupling of energy costs by 2000.[30]

How do these estimates measure up to the recent history of actual energy price increases? In the past generation world energy prices increased 11 times.[31] In the first half of the present decade the price of Arabian oil increased 7.6 times.[32] In the four years from 1972 to 1976 the price of uranium increased eightfold.[33] In the four years from 1973 to 1977 the average wellhead price of natural gas in the U.S. more than quadrupled.[34] Linear extrapolations of past energy trends[35] yield a projected price increase for oil and natural gas from 10- to 100-fold by 2000.

We cannot predict the future price of energy with certainty; informed estimates range from a doubling in price by 2000 to a 100-fold increase. But the recent trends cited above give strong indication that the average price of energy will increase exponentially in the future and that the conservative estimates cited above are unrealistically low. What seems certain is that the price of energy will increase faster than the price of non-energy-related prices, which may in turn force a fundamental revision of the price structure of industrial economies. This change in price structure underlies the view of the future that is envisioned in the second half of this book. For now, the main point is that the transition of industrial civilization does not require the depletion of energy; a relative increase in the price of energy is a sufficient driving force.

The Limits of Technology

Can technology rescue industrialism from the energy price squeeze? As we have observed, each time industrial society has in the past exhausted one source of fuel, such as firewood, technology has helped humanity to harness yet another source of energy, such as fossil fuels.[36] With such an impressive record of past success it is not surprising that some people retain faith in the power of technology to extricate industrialism once again.

Perhaps the ultimate proponent of the technological "fix" is Herman
Kahn, who asserts that with "current technology we can support a worldwide population of twenty billion people with twenty thousand dollars per
capita . . . in twenty-five years you can fix every problem in America—
pollution, growth, traffic, housing."[37] Kahn is not alone in this view; other
experts expect that technology will eventually enable us to mine common
rock economically for its mineral content:

> It is simply not true, as is often remarked, that average rock will never
> be mined. . . . Consider the record of the past century, during which
> mineral supply has kept well ahead of long-term growth in mineral con-
> sumption. Why should a process of technological development that has
> been operating so well come suddenly to an end? New development will
> of course have to come more rapidly with higher rates of consumption,
> but it can be argued that the steepest exponential curve of all is the
> growth of human knowledge.[38]

Why indeed should technological development come to an end? The
answer lies in the nature of technology and its relation to industrialism.
Technology is the application of knowledge to the solution of practical
problems; it is thus based on the production of material knowledge
through scientific research. As mentioned when we discussed the service
industries (Chapter 5), research is heavily dependent on the availability of
capital that is in turn generated by the basic industries. Thus industrialism gives rise to knowledge and its application, technology, which in
turn furthers the cause of industrialism in a mutually reinforcing or positive-feedback relationship (Figure 8.4). But when the resource cycle is
slowed because of energy and resource scarcity, capital availability declines; and as we have seen, the service industries in general and research
in particular are especially vulnerable to shortages of capital. From 1970
to 1976 federal expenditures for basic research in the U.S. declined in
real terms by 22%, and the trend is expected to continue in the coming
decade.[39] As a result of capital shortages, academic research in the U.S. is
visibly deteriorating,[40] and the basic industries such as steel, glass, and
wood are increasingly resistant to technological innovation.[41]

Even in the absence of capital constraints the growth of new knowledge on which technology is based is subject to an intrinsic limit. The history of science amply documents that when a given branch of knowledge
is first established as a discipline, scientific "breakthroughs" are frequent.
But as the discipline matures, fewer breakthroughs occur because the
number of possibilities declines steadily. Knowledge, and thus technology, is subject to the same depletion dynamic as fossil fuels and Easter

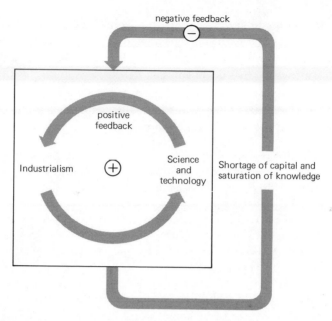

8.4   Feedbacks in the relation between industrialism and technology. See text for description.

eggs. What is easily discovered is found first, and what remains is increasingly harder (and therefore more expensive) to find. It is typical of science that as a discipline matures, its practitioners necessarily turn their attention away from producing new ideas and toward the refinement of existing ideas.

The practical impact of such saturation of the knowledge curve is well illustrated in the agricultural industry. Owing to the absence of new scientific breakthroughs and the corresponding lack of technological advance, the corn yield on experimental farms in the U.S. has remained static for over a decade.[42] Agricultural research has pushed photosynthesis to its technical limits. As a result, agricultural scientists now believe that industrial agriculture is being sustained on past breakthroughs, and faces a scientific dead-end for the future.[43]

As for the future, no one can reasonably deny the possibility of important technical breakthroughs, especially in the area of energy development where research funds are likely to be concentrated increasingly. Some of these technological developments will undoubtedly prolong industrialism. But technology is the servant of industrialism, not its savior. In times of economic growth, as have prevailed during the past two centuries, it is easy to transpose the two roles. In times of economic decline,

however, technology is among the first of human enterprises to be afflicted. A limited form of technology is likely to serve the interests of human
survival beyond the Industrial Age. But writers who base their hope for
the continuation of industrialism on the steeply exponential growth of
knowledge do not take into account that the knowledge curve ultimately
owes its shape to the energy curve, and the energy curve must soon peak.

Limits
and the Principle
of Interrelatedness

Until now we have considered the various limits to industrialism as if each
were isolated from all others. In reality, everything interacts with everything else, whether the parts of a watch, the organs of the human body,
organisms in the food chain, or the basic industries of the resource cycle.
To consider each limit to industrialism in isolation is a necessary and
useful simplification, but it is artificial because all such limits are interrelated and operate simultaneously.

The First-Order Interrelation

In order to appreciate how the limits to industrialism interact, let us
begin with the simplest possible interaction, namely, that which occurs
between two limits, and which I will term a first-order interrelation. By
dealing with only two limits we are still indulging in the artificial, but the
exercise will prepare us for the more realistic and complex examples to
follow.

Consider two major limits to industrialism that we have discussed at
length, the supply of energy and the supply of mineral resources. As we
have seen from discussion of the resource cycle, energy and mineral
resources are inextricably linked; energy is required to extract and process minerals, and minerals are in turn essential to construct machinery
that extracts and processes energy. In 1974 the recovery and processing of
industrial materials consumed fully 16% of the total U.S. energy
budget.[44] We have also seen that as high-quality deposits of minerals are
depleted, the energy cost of extracting lower grade ores grows exponentially (Chapter 7, Figure 7.7). In other words, a small reduction in the
quality of ore demands a large increase in the amount of energy needed to
process it. As a result of this relationship U.S. mineral production in the

past 50 years rose 50%, and yet the energy required to produce those minerals increased 600%.[45] When we exhaust the available high-grade metal ore in the foreseeable future, we will have to move to lower grade ores; but because we are also depleting high-grade energy sources, it will be physically impossible to invest exponentially increasing quantities of energy into the extraction of metal from lower grade ore. The two limits collide to create an even more imposing limit than might be imagined from either one alone.

The Second-Order Interrelation

Now let us progress to a slightly more realistic but still simplified example of three interacting limits: energy, food, and water. As discussed in Chapter 4, oil shales in the western United States offer some promise of a domestically available supplement to declining oil and natural gas. In 1974, prior to the three-year drought that gripped the western states, the Bureau of Land Management of the U.S. government completed a report analyzing the water requirements for extracting oil from the shale deposits.[46] The report concluded that unless agricultural activity were dramatically curtailed, there is insufficient water in the western states to pursue even modest pilot extraction projects. Moreover, the proposed shale operations would increase the salinity of water in the lower Colorado River, with attendant threats to the productivity of the rich agricultural valleys of southern California, the southwest U.S., and northern Mexico.

This brief example of a second-order interrelation illustrates the consequences of considering three interacting limits simultaneously, in this case energy, water, and food production. Considered in isolation the oil shales represent an inviting additional source of energy; but considered in the context of only two other limits, they lose some of their appeal.

The Nth-Order Interrelation

In the "real" world, first- and second-order interrelations do not exist. Instead, the many individual limits that we have identified operate simultaneously and interactively. When mathematicians are faced with a large number of unknown magnitude, they typically call it n. To borrow their convention, every interaction between the many limits to industrialism is an nth-order interrelation. Of course it is impossible to outline fully an nth-order interrelation; the numbers are too large and the variables too complex. It is nevertheless instructive for the sake of illustration to sketch the beginnings of an nth-order interaction, using the technique of scenario.

Let us suppose that the State of California is forced by the shortage of oil to meet its growing energy needs with coal. Coal is a dirty fuel: its combustion yields fly ash, which causes genetic mutations in bacteria and therefore may pose a cancer threat[47] (limit #1). Moreover, burning coal adds CO_2 to the atmosphere, threatening climatic disruption by means we discussed earlier (limit #2). Coal also faces transportation constraints (limit #3) and unanticipated cost escalations owing to dissatisfaction in the mining industry[48] (limit #4), not to mention conplex federal regulatory problems (limit #5) that derive in part from the complicated pattern of ownership of coal rights (limit #6). Finally, the combustion of coal generates air pollution, which, as we have discussed, impairs agricultural productivity (limit #7). Agriculture is the basis of California's economy; in our hypothetical scenario, the threat of disrupting agriculture that is posed by coal combines with the other limits to compel a switch to nuclear power.

Nuclear power, however, is constrained by a related set of interacting limits. To begin, nuclear power plants require vast quantities of water for cooling (limit #1). Additional nuclear power plants cannot be sited on the coast, however, because environmental concerns (limit #2) and seismic instability (limit #3) have catalyzed insurmountable legal opposition (limit #4). So, there is no choice but to locate the nuclear facilities inland, in the central valleys. But here the available water is already insufficient to meet agricultural requirements (limit #5). Thus the nuclear plants must use dry cooling towers. These towers, which stand some 50 stories tall, are much less efficient than water-cooled towers, resulting in a substantial increase in "waste" heat (limit #6). The increased thermal pollution generates "heat islands" of the kind that already tower miles above such major urban centers as Chicago and New York. Such heat islands divert rainstorms to downwind locations; hence they reduce rainfall in the already dry valleys, leading to a decline in agriculture productivity, catalyzing legal opposition which forces . . .

We could continue the scenario, but the point is probably made that the nth-order interrelation is beyond the assimilative powers of the human mind. Indeed, the number of possible combinations quickly swamps the memory of even the largest computer. The nth-order interrelation that represents the immediate limit to industrialism is like an elusive, shimmering web. Perturbation of a single strand generates standing waves everywhere at once, and then the waves themselves interact to produce an entirely new set of conditions. The behavior of this metaphorical web is an "emergent" property that cannot be deduced from the properties of the individual strands, just as the behavior of water would be difficult to predict knowing only the properties of hydrogen and oxygen. Moreover, the nth-order interrelation involves parallel events that are not amenable to the linear logic to which the Western mind is accustomed and the written word is suited.

It is largely for these reasons that no individual or group, regardless of how well trained and financed, can formulate an exact timetable for the future of industrialism. Energy and resource depletion would appear to set an outside limit of one to several decades, but within this broad time frame anything is possible at any time. A prolonged drought or several severe winters could trigger the transformation with a few years; or sudden, unexpected breakthroughs in fusion power could prolong industrialism, in which case an entirely new set of limits would reign. The complexity of the possible interactions counsels against the proposal of specific timetables. The interrelatedness of limits and their effects suggests rather that the transformation may strike unexpectedly, like the biblical "thief in the night." At the very least, it is likely that the cost of both energy and resources will increase disproportionately between now and the year 2000. This economic force is alone sufficient to the view of the future that we are now in a position to develop.

NOTES

1. E.g., W. J. Chancellor and J. R. Gross, "Balancing Energy and Food Production, 1975–2000," *Science* 192 (1976): 213–18.
2. E.g., E. J. Kormondy, *Concepts of Ecology*, 2nd ed. (Englewood Cliffs, N.J.: Prentice-Hall, 1976).
3. E.g., L. Stark, *Neurological Control Systems: Studies in Bioengineering* (New York: Plenum, 1968); J. H. Milsum, *Biological Control Systems Analysis* (New York: McGraw-Hill, 1966); L. E. Bayliss, *Living Control Systems* (San Francisco: W. H. Freeman, 1966).
4. C. A. Berg, "Process Innovation and Changes in Industrial Energy Use," *Science* 199 (1978): 608–14.
5. J. P. Martino, *Technological Forecasting for Decisionmaking* (New York: American Elsevier, 1972), p. 447.
6. J. Vinocur, "Outlook Held Bleak for Manufacturers of Atomic Reactors," *New York Times*, 21 September 1977, p. A1; "The Nuclear Boom Is Going Bust," *San Francisco Chronicle/Examiner*, 26 June 1977, p. 1.
7. V. K. McElheny, "Solar Energy Future: Optimism Is Restrained," *New York Times*, 31 December 1976, p. D3.
8. A. B. Lovins, *World Energy Strategies* (Cambridge, Mass.: Ballinger, 1975), p. 22.

9. E.g., D. B. Brooks and P. W. Andrews, "Mineral Resources, Economic Growth, and World Population," in *Materials: Renewable and Non-renewable Resources*, edited by P. H. Abelson and A. L. Hammond (Washington, D.C.: American Association for the Advancement of Science, 1976), pp. 41–47.

10. R. L. Gordon, "The Hobbling of Coal: Policy and Regulatory Uncertainties," *Science* 200 (1978): 153–58.

11. S. Rattner, "Economy Unharmed by Mine Strike; Coal Industry Faces Investment Lag," *New York Times*, 16 August 1976, p. 45.

12. W. J. Mead, "An Economic Appraisal of President Carter's Energy Program," *Science* 197 (1977): 340–45.

13. M. A. Adelman, "Efficiency of Resource Use in Crude Petroleum," *Southern Economic Journal* 31 (1964): 101–22.

14. Ibid.; Mead, "An Economic Appraisal of President Carter's Energy Program"; R. F. Naill and G. A. Backus, "Evaluating the National Energy Plan," *Technological Review* 79 (1977): 3–7.

15. P. Barnes, *The Sharing of Land and Resources in America* (Washington, D.C.: The New Republic, 1972).

16. E.g., A. S. Luchins and E. H. Luchins, *Rigidity of Behavior* (Eugene, Ore.: University of Oregon Press, 1959).

17. S. Freud, *Collected Papers*, edited by J. Strachey (London: Hogarth, 1953), 4:391.

18. E.g., R. L. Heilbroner, *The Human Prospect* (New York: W. W. Norton, 1975).

19. E. T. Hayes, "Energy Implications of Materials Processing," in *Materials: Renewable and Non-renewable Resources*, edited by P. H. Abelson and A. L. Hammond (Washington, D.C.: American Association for the Advancement of Science, 1976), pp. 33–37 (quoted information from p. 34).

20. Lovins, *World Energy Strategies*, p. 51.

21. National Petroleum Council, *U.S. Energy Outlook—An Initial Appraisal, 1971–1985* (Washington, D.C.: National Petroleum Council, 1971), vol. 1.

22. J. E. Hass, E. J. Mitchell, and B. K. Stone with the assistance of D. H. Downes, *Financing the Energy Industry* (Cambridge, Mass.: Ballinger, 1974).

23. D. J. Rose, "Nuclear Electric Power," *Science* 184 (1974): 351–59.

24. Hass et al., *Financing the Energy Industry*.

25. Ibid., p. 6.

26. National Academy of Engineering, *U.S. Energy Prospects* (Washington, D.C.: National Academy of Sciences, 1974).

27. S. D. Freeman, *Energy: The New Era* (New York: Vintage, 1974), p. 35.

28. "Big Rate Increase for Natural Gas Approved by F.P.C.," *New York Times*, 28 July 1976, p. 1; S. Rattner, "F.P.C. Cuts Gas Price Rise by 25%; Admits Error in July Estimates," *New York Times*, 21 October 1976, p. 1.

29. A. L. Hammond, W. D. Metz, and T. H. Maugh III, *Energy and the Future* (Washington, D.C.: American Association for the Advancement of Science, 1973), p. 139.

30. E. Cowan, "Solar Heat Competitive with Electric, Agency Finds," *New York Times*, 30 December 1976, p. 1.

31. C. T. Rand, "The World Oil Industry," *Environment* 15 (1976): 13–26.

32. S. Rattner, "Increased Oil Prices Held Unlikely to Curb Recovery," *New York Times*, 3 November 1976, p. 61.

33. P. Hofmann, "Outlook for a Rise in the Price of Oil Pushes Uranium Up," *New York Times*, 26 November 1976, p. D1.

34. S. Rattner, "16% Rise Last Winter Makes New York Gas Costliest," *New York Times*, 9 July 1977, p. C23.

35. Foster Associates, *Energy Prices, 1960–1973*, Energy Policy Project of the Ford Foundation (Cambridge, Mass.: Ballinger, 1974).

36. Berg, "Process Innovation and Changes in Industrial Energy Use."

37. H. Kahn, quoted in R. Bundy, *Images of the Future* (Buffalo, N.Y.: Prometheus, 1976).

38. Brooks and Andrews, "Mineral Resources . . . ," p. 42.

39. P. M. Boffey, "Carter Aides Lament Research Decline," *Science* 197 (1977): 32.

40. J. Walsh, "The State of Academic Science: Concern about the Vital Signs," *Science* 196 (1977): 1184–85.

41. F. P. Huddle, "The Evolving National Policy for Materials," in *Materials: Renewable and Non-renewable Resources*, edited by P. H. Abelson and A. L. Hammond (Washington, D.C.: American Association for the Advancement of Science, 1976), pp. 18–23.

42. L. M. Thompson, "Weather Variability, Climatic Change and Grain Production," in *Food: Politics, Economics, Nutrition and Research*, edited by P. H. Abelson (Washington, D.C.: American Association for the Advancement of Science, 1975), pp. 43–49.

43. V. K. McElheny, "Scientific Dead-end Faces U.S. in Its Agricultural Technology," *New York Times*, 3 May 1976, p. 1.

44. R. S. Claasen, "Materials for Advanced Energy Technologies," *Science* 191 (1976): 739–45.

45. T. S. Lovering, "Mineral Resources from the Land," in *Resources and Man* (San Francisco: W. H. Freeman, 1969), pp. 109–34.

46. J. McCaull, "Wringing out the West," *Environment* 16 (1974): 10–17. See also J. Harte and M. El-Glasseir, "Energy and Water," *Science* 199 (1978): 623–43.

47. C. E. Chrisp, G. L. Fisher, and J. E. Lammert, "Mutagenicity of Filtrates from Respirable Coal Fly Ash," *Science* 199 (1978): 73–75.

48. A. J. Parisi, "Strike End Threatens Coal Upheaval," *New York Times*, 28 March 1978, p. 53.

PART II
CONSEQUENCES

In the preceding chapters I have summarized the evidence that industrial civilization stands on the threshold of an epochal discontinuity, occasioned by the depletion of energy and natural resources. The thesis is not new, and it is certainly arguable, as is any thesis bearing upon the future. At the very least, I hope the reader is persuaded that energy and resource depletion within the foreseeable future is a feasible possibility, and one that therefore ought now to be seriously addressed by individuals and governments alike.

For the remainder of this book I will consider the depletion of energy and resources and the corresponding increase in their price as a "given," and explore some of the consequences for industrial civilization. If the thesis is proven wrong by the passage of time, then the chapters that follow will serve mainly to emphasize the dependence of our present way of life on energy and natural resources. If the thesis is proved correct, however, then the chapters that follow may sketch a way of life toward which industrial civilization will necessarily evolve in the coming decades; and in doing so they may suggest preparations that we can usufully entertain as individuals and societies.

Chapter 9
THE STAFF
OF LIFE

Because everything in industrial civilization is shaped by energy and resources, their depletion is correspondingly certain to transform every facet of life. And because everything is interrelated with everything else, the transformation may be expected to occur in an infinite chain of simultaneous events that begins everywhere at once and ends nowhere in particular. Industrial societies may be likened to a watch, or a human body, in which the proper function of the whole depends upon the tightly orchestrated operation of the individual parts. Owing to this interdependence, changes in any one component can have rapid and often unforeseen effects throughout the entire system.

And yet, the principle of interrelatedness notwithstanding, certain components of any system are more proximate to the operation of the whole, and more immediate in effect, than other components. A watch has its mainspring, the human body its heart. The heart and mainspring of industrial society—or for that matter any human society—are its food production system. When oil wells can no longer deliver net energy we will presumably manage without jet airliners and climate-controlled shopping centers, and even without automobiles; but we cannot survive without an effective system of food production. In short, energy and resource scarcity may furnish the fulcrum of change, but our need for food may act as the immediate lever.

We therefore begin our exploration of the consequences of energy and resource scarcity with the subject of food. In this chapter I will briefly trace the evolution and present structure of the U.S. food system, with the intent of showing how fully our supply of food has been shaped by, and is now dependent on, fossil fuels and natural resources. The exercise will hopefully heighten appreciation of our daily bread; it will ennumerate the ecological imbalances of industrial agriculture; and it will help us to fathom the depth of the changes that may soon envelop us.

Roots:
The Agricultural Revolution

The acceleration of the food cycle that underlies the Industrial Age has been aptly termed the Agricultural Revolution.[1] Like the Industrial Revolution with which it was linked, the Agricultural Revolution was powered by fossil fuels and punctuated by technical innovation. Thus the first agricultural tools brought to the New World by the Pilgrims were nearly identical to those used by the ancient Egyptians.[2] But in the short span of a century the steam engine was invented, the cotton gin developed, and the wooden plow was surperseded by the cast-iron plow and then again

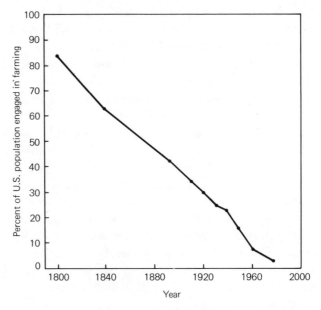

9.1 Farm population of the United States over the past century and a half, expressed as a percentage of the total population. (data from L. Tweeten, *Foundations of Farm Policy* [Lincoln, Neb.: Univ. of Nebraska Press, 1970], p. 126, except for the 1976 point, which is from J. S. Steinhart and C. E. Steinhart, "Energy Use in the U.S. Food System," *Science* 184 [1974]: 307–16)

by the steel plow.[3] In the half century from 1830 to 1880 advances in farming technology reduced the time required to harvest an acre of wheat from 75 hours to 13 hours.[4] By the end of the 19th century a single farmer could produce enough food for three people. The excess labor that was thereby freed from food production was channeled into manufacturing, setting the stage for the accelerated positive feedback with which we are by now familiar.

In its earliest stages, U.S. agriculture was a decentralized, subsistence operation. As is typical of nonindustrial societies, the vast majority of U.S. citizens were farmers (Figure 9.1). Food was eaten where it was grown because there was no way to transport it elsewhere. But the laying of the U.S. railroad system in the 19th century marked the beginning of a new era in U.S. agriculture. Now for the first time farm products could be transported cheaply and efficiently to distant markets. By the middle of the 19th century the Ohio vegetable farmer could drink Wisconsin milk and bake bread from Kansas wheat. Thus did the "iron horse" free the farmer from the bondage of subsistence. With farmers no longer dependent on themselves for all their needs, the American farm began its long march toward specialization.[5]

The Capital Phase of U.S. Agriculture

The railroads not only opened the urban market to farm products, they also opened the rural market to the products of the factory. Now even the poorest farmer could afford a steel plow, and the railroad could deliver it in a matter of weeks. Carried on the back of the iron horse, U.S. agriculture raced away from subsistence farming and toward the 20th century. During this era of farm capitalization, farm labor was steadily replaced by farm machinery, with the result that the productivity of the individual farmer grew. Thus the output of U.S. farms climbed steadily throughout the 19th century (Figure 9.2) despite the steady shrinkage of farm population (Figure 9.1).

By 1920, U.S. farm output showed signs of approaching a plateau (Figure 9.2). But the exuberant industrial sector of the U.S. economy, carried on a growing floodtide of fossil fuels, crowned the capital era of U.S. agriculture with a major new event. Animal power, long the mainstay of the American farm, was replaced by the mechanized power of tractors and trucks. The effects on farm output were immediate. Not only was

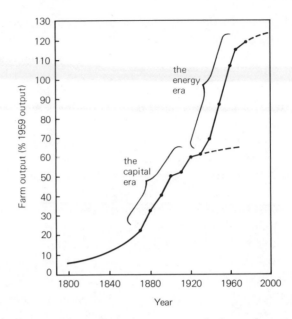

9.2   The output of U.S. farms over the past century. Note the latent plateau in the 1920s (at the end of the capital era), followed by accelerated growth during the energy era (1940 to present). The curve is extrapolated backward from 1870. (data from L. Tweeten, *Foundations of Farm Policy* [Lincoln, Neb.: Univ. of Nebraska Press, 1970], p. 129, except for the most recent point, which is from J. S. Steinhart and C. E. Steinhart, "Energy Use in the U.S. Food System," *Science* 184 [1974]: 307–16)

the productivity of each farmer magnified; in addition, 86 million acres of prime arable land—fully one-fourth of the U.S. total—was freed from the production of animal fodder. The plateau in farm output that appeared to be developing in the 1920s was converted overnight into a fresh spurt of exponential growth. By 1940 each farmer could feed five people, and the percentage of the U.S. population working on farms continued its unbroken descent.

The Energy Phase of U.S. Agriculture

During the capital phase of U.S. agriculture, farm output increased largely because capital inputs to farms also increased. Thus in the half century from 1880 to 1930 the output of U.S. farms expanded by 97%, while production inputs (land and machinery) expanded by 83%.[6] In sharp contrast, the 35 years from 1930 to 1965 saw another near doubling of farm output, but during these same years capital inputs hardly changed. From 1930 to 1965 the output of U.S. farms increased by 88%, while production inputs increased only 6%.[7] Thus the increased farm output during recent decades had a different cause—energy. From 1940 to 1970, the energy input to the U.S. food system more than quadrupled (Figure 9.3). Beginning with World War II, the burden of transporting farm products switched from railroads to energy-intensive trucks; horses and mules continued to be replaced by gasoline-driven tractors; and most significant, the use of fertilizers—manufactured directly from fossil fuels—grew exponentially[8] (Figure 9.4).

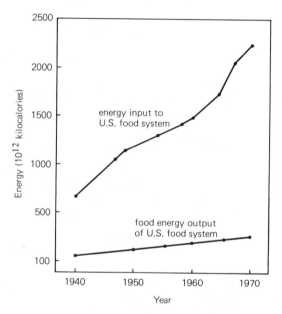

9.3 Energy input and output for the U.S. food system. The energy input has risen fourfold since 1940, while the energy output (equivalent to food energy consumed) has less than doubled. (from J. S. Steinhart and C. E. Steinhart, "Energy Use in the U.S. Food System," *Science* 184 [1974]: 307–16, Figure 1)

The effects of fertilizers on U.S. farm output are difficult to exaggerate. Prior to 1940, grain productivity was about the same as in any traditional agriculture. But after 1940 fertilizer application to U.S. croplands grew exponentially, and grain production followed suit (Figure 9.4). Of course the increased grain yield had multiple causes: unusually favorable weather, new cultivating techniques, and new genetic strains all contributed, as we shall discuss. But it is estimated that about half of the increased grain yield since 1940 has resulted directly from the increased use of fertilizer.[9] American farmers now use more energy per acre in fertilizer than in tractor fuel.[10]

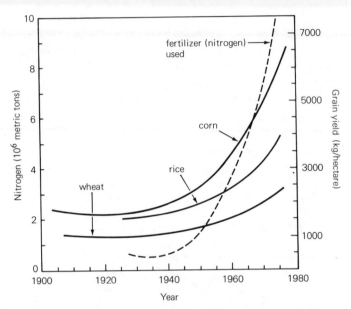

9.4 Fertilizer use and grain yields in the United States. Beginning in World War II, fertilizer use has increased exponentially, and grain yields have followed suit. (from W. J. Chancellor and J. R. Gross, "Balancing Energy and Food Production, 1975–2000," *Science* 192 [1976]: 213–18, Figure 7)

Farm Methodology

The heavy reliance of U.S. agriculture on energy has radically altered our food-production system. Plowing, planting, cultivating, and harvesting that were once accomplished with muscle, animal power, and simple agricultural implements are now performed with towering machines propelled by fossil fuels (Figure 9.5) and energy-intensive chemicals. Corn, for example, is planted with enormous tractor-drawn devices that mechanically insert the seeds in perfect rows, precisely one inch beneath the surface of the soil and eight inches apart. The seeds are counted and their release monitored by a computer inside the air-conditioned cab. The same machines simultaneously spread herbicides between the freshly planted rows to suppress the growth of weeds chemically. The farmers, who obtain most of their information from the chemical companies that manufacture the herbicides, typically believe that the chemicals affect the weeds but not the corn. But geneticists have found that atrazine, the most widely used herbicide, is incorporated into the corn plant and transformed into a potent mutagen, or potential cancer-causing substance.[11] Almost 100 million pounds of atrazine have been spread across the country;[12] the majority of corn crops are treated with this substance. Whether this herbicide, or the mutagens it induces, have entered the food chain in beef and milk from corn-fed cattle is unknown.

With the prompting of chemical companies, farmers are beginning now to use chemicals even in place of the plow, with the so-called no-till

9.5 Combines harvesting wheat in the state of Washington. (from P. R. Day, ed., *The Genetic Basis of Epidemics in Agriculture* [New York: NYAS, 1977]; courtesy Grant-Heilman)

farming method.[13] The key to no-till farming is an especially poisonous herbicide called paraquat, marketed exclusively by the Cheveron Chemical Company. Paraquat is sprayed on an unplowed field; three days later, the soil is devoid of weeds, loose and powdery, and ready to plant. The long-term effects of paraquat on the land, crops, and animals that eat the crops, are unknown. According to the U.S. Department of Agriculture, fully half the nation's croplands could be farmed the no-till way by the year 2000.[14]

Farm Population

The driving force behind the transformation of farm methodology is economic. A single barrel of oil contains the energy equivalent of 6,000 hours of human labor (calculated from data in G. Leach).[15] At $15.00 per barrel, the "energy slaves" bound up in oil work for less than a penny per hour. Thus in the U.S. today, as in all industrialized nations, fossil fuels have displaced human labor from agriculture (Figure 9.6), and farm population has declined accordingly (Figure 9.1). With fossil fuels, one U.S. farmer can now plant, cultivate, and harvest 500–1,000 acres of corn each year.[16] With fossil fuels, one farmer can now grow enough food for 50 people, and less than 3% of the U.S. population can feed this nation and several others. Because fossil fuels have been abundant and cheap, food has been likewise abundant and cheap, and is now taken for granted throughout most of the industrialized world. Even children's nursery rhymes reflect this consciousness:

> And when the city children
> Run out of plum and prune
> They know more trucks are coming
> As surely as the moon.[17]

Farm Size

Prior to the industrialization of U.S. agriculture the prospective farm family needed little more than strong backs, strong wills, land, and a mule. But fossil fuels have added new prerequisites. The high expense of machinery and land now requires that the prospective farmer also have cash,

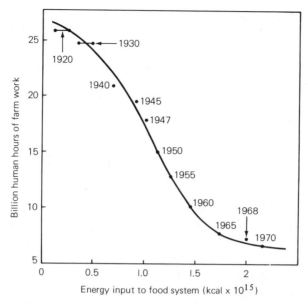

9.6 The relation between energy use and human labor in the U.S. food system. As energy use has increased, human labor has declined steadily. (from J. S. Steinhart and C. E. Steinhart, "Energy Use in the U.S. Food System," *Science* 184 [1974]: 307–16, Figure 3)

credit, and an ability to manage money. Because large farms are generally worth more, a farmer's access to cash and credit increases with farm size, giving the large farm an important economic advantage. Thus as fossil fuels have come to dominate American farming, small farmers have been squeezed out of business and their lands swallowed up by their larger and richer neighbors. Farm size has therefore directly paralleled energy use. In the 25 years from 1940 to 1965, the average American farm doubled in size, from 167 to 328 acres.[18] In California, the most important agricultural state, the trend is even more pronounced: from 1930 to 1969 the average California farm nearly tripled in size, from 224 to 627 acres.[19]

Corporate Agribusiness

Individual families find it increasingly difficult to raise the capital that is essential to modern farming. Large corporations, in contrast, have ready access to capital, and they are investing it increasingly in agriculture. As a consequence, the U.S. family farm—once the centerpiece of the American heritage—is being replaced by the U.S. corporate farm. In California, an estimated one-fourth to one-third of all cropland is owned and operated by multinational corporate giants like Standard Oil and IT&T.[20] Nationwide, the percentage of cropland owned by corporate farms is estimated variously from 1%–5%[21] to 10%.[22]

What is the impact of the corporate farm on U.S. agriculture? Corporations are governed by the law of economic survival; those that flourish are by definition those that reap the greatest return from the least investment. Therefore, in contrast to the family farm of yesteryear, corporate farms have little stake in nurturing the soil for the sons and daughters that will someday inherit the earth. Instead the corporate farm is obliged by economics to squeeze the earth for maximum short-term profit. Family farms must compete in the same marketplace as the corporate giants. The family farm, handicapped already by its smaller capacity to absorb the periodic financial losses that are typical of agriculture, and bombarded by propaganda from the chemical industries,[23] is sucked into the same economic vortex as the corporate farm. Farmers can hardly be faulted for struggling to survive. But the principal and immediate loser in this contest is not the farmer—it is the earth and its future inhabitants. By its heavy reliance on chemical farming, including pesticides (see below), industrial agriculture may be rendering our soils unfit for future use.

The Evolution of the Supercrop

Of the many ways to maximize farm profit, one of the surest is to grow only those crops that yield the highest financial return per unit of land, labor, and capital input. Driven by such economic incentive, modern agricultural technology has manipulated the genetic makeup of food crops, culminating in high-yielding varieties called "supercrops."

What Makes Supercrops Super?

The scientific basis of supercrops was laid in the 19th century by an obscure Austrian monk, Gregor Mendel. By selecting and crossbreeding sweetpeas of the same color, Mendel discovered that the color could be emphasized in subsequent generations. Molecular biologists have since found that the color of the sweetpea flower, and for that matter all inherited traits in plants and animals alike, are coded by molecules of nucleic acid packaged in genes. Geneticists cannot yet manufacture new genes, but they can produce new combinations of existing genes by selective breeding of plants having the desired traits.

What characteristics have agricultural technologists deemed desirable in selectively breeding the supercrops? By far the most important trait is responsiveness to fertilizer. The original native crop plants that fed our ancestors, called the "land races," were required by natural selection to meet most of their own nutritional needs, and they are relatively indifferent to fertilizers. But the supercrops have been prompted by unnatural selection to respond to heavy doses of fertilizer with a doubling or tripling of yield (Figure 9.7). Likewise, the supercrops have been bred for responsiveness to regular and copious irrigation. In other words, the supercrops are "super" only in the narrow sense that they respond optimally to the specialized and artificial conditions of modern industrial agriculture.

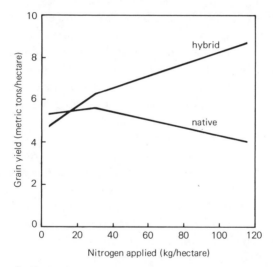

9.7 Responses to fertilizer of two varieties of rice, a native plant and a modern hybrid "supercrop." Without fertilizer, the native plant is more productive. When fertilizer is applied, however, the tall stalks of the native plant bend and break, reducing the yield. The supercrop hybrid is a "modern dwarf," bred for short, stiff stalks that hold the grain aloft until harvested. (from P. R. Jennings, "The Amplification of Agricultural Production," in *Food and Agriculture* [San Francisco: W. H. Freeman, 1976], p. 125–33).

The Vanishing Genetic Resource

As long as fertilizer and water can be applied with abandon, supercrops produce more food than do the land races from which they were bred. But humankind has paid a steep price for this extra productivity. In the process of breeding the supercrops many of the genes that impart resistance to plant disease have been unavoidably lost.[24] Moreover, the supercrops are by design genetically homogeneous: in a field of corn, for example, each cornstalk has exactly the same genetic makeup as every other, like identical twins. Thus if one plant in a field is especially susceptible to a particular disease, the whole field is vulnerable.[25] The supercrops are, in short, genetic freaks that could never have evolved without human intervention. They are like cattle bred for no legs and no horns, maximally productive of meat under artificial feedlot conditions, but otherwise helpless. But because fossil fuels have been plentiful and cheap, supercrops have been maximally profitable and have largely replaced the original land races as the source of human food. Throughout human history some 3,000 species of plants have been cultivated for food; today the human population is fed largely by only 15 plant species.[26]

The land races, staple of our ancestors, are perfectly suited by evolution to the growing conditions that have prevailed throughout most of human history. They are hardy and tough; they can survive flood and drought; and because each individual plant contains a unique combination of genes, the land races possess a wide variety of genetic tools for combating diseases. But in many cases their unique combinations of genes no longer exist. Like the passenger pigeon, many land races are for practical purposes extinct.

Genetic variability is disappearing also from domestic animals, as small numbers of specialized livestock breeds are replacing traditional stocks. A 1975 U.N. report identified 115 breeds of cattle in Europe and the Mediterranean Basin alone as "in a relic state, in danger now or in danger in the future."[27] According to the report, Europeans will soon obtain all their milk from only two or three specialized dairy breeds. Extinction of native stock will mean loss of yet another genetic resource that could prove important to human survival in an energy- and resource-scarce future.

The Age of the Monocrop

The specialization of farms that was spurred originally by the railroads has been accelerated by the coming of the supercrop. Farmers can now derive maximum profit by planting a single supercrop and harvesting it with machinery that is typically specialized for that crop alone (Figure 9.5). Accordingly, most U.S. farms now produce but one crop, a practice known as monocropping. Planting farmlands in the same crops year after year is indeed profitable in the short term, but the practice drains the soil of its nutrients. The chemical fertilizers intended to replace the depleted nutrients cannot restore organic matter to the soil,[28] and hence farmlands become powdery and vulnerable to wind and water erosion. Early experience with no-till chemical farming suggests it may harden the earth to the degree that not even a plow can penetrate it. At least one farmer has actually resorted to dynamite to break up the land following several seasons of no-till farming.[29] "Give," we are advised, "and ye shall receive"; with industrial farming practices we have reaped the bounty of the earth but not given back in kind. If we continue to take from the soil more than we return, the soil may lose its capacity to give anything.

The Ecology of Monocropping

Monocropping not only drains croplands of ther fertility; it also violates one of nature's most fundamental laws, the laws of species diversity. Natural ecosystems are typically composed of a rich diversity of plant and animal species. When any one species multiplies rapidly and achieves temporary dominion, its predators also multiply owing to their increased food supply and the original equilibrium is automatically reestablished. Thus does diversity assure balance between the species, and the consequent stability of the ecosystem.[30]

Modern industrial agriculture ignores this eco-logic when it plants vast land areas in a single crop. Amber waves of grain that stretch from one horizon to the other may be beautiful to behold (Figure 9.5), but they comprise a disastrously simplified and unstable ecological community. Such monocultures guarantee a virtually unlimited feeding and breeding ground for the plant pathogens—insects, viruses, and fungi—that are best suited to that crop. Nature strives to correct the imbalance with periodic epidemics, such as the southern corn leaf blight that decimated the U.S. corn crop in 1970.[31] Because supercrops are genetically homogeneous, they are especially vulnerable to such epidemics.[32] Industrial agriculture has responded to such epidemics not by working with nature to restore balance, but rather by working against nature to preserve imbalance. Instead of rotating and intermixing crops to minimize their overall vulnerability to pathogens, modern industrial agriculture has opted to protect its monocrops with pesticides.

Pesticides:
A Case Study
in Diminishing Returns

Without massive and frequent doses of pesticides, monocropping would be difficult, and it would certainly be less productive. But nature abhors the imbalance represented by monocultures, and she resists with the same device that underlies evolution—genetic diversity. The wild populations of pathogens that attack monocrops are like the land races in that their genetic heritage is extraordinarily diverse. This diversity guarantees that a small fraction of a given pathogen population is endowed with exactly that combination of genes that make it resistant to a given pesticide. The resistance might manifest itself as an impervious body wall that denies entrance to the pesticide, or an especially rapid and thorough metabolic breakdown of the pesticide inside the target organism. Whatever the cause of the resistance, the effect is that a small fraction of the pathogen population survives. This resistant strain now finds itself in the midst of abundant food with no competition from its dead brethren. Under such optimal conditions, the resistant strain flourishes and the population multiplies exponentially. Our reliance on pesticides has thus cast us in the role of evolution; when we apply pesticides to genetically diverse pathogen populations, we act as a force of natural selection and accelerate the evolution of resistant strains that are indifferent to our poisons.

In the past, industrial agriculture has employed fancy technological footwork to keep one step ahead of the resistant strains it produces. As quickly as new strains arise, the pesticide industry has been able to develop yet a new poison, or geneticists have bred yet another supercrop that is not so vulnerable to existing resistant strains. But industrial agriculture is beginning to lose this race with nature; resistant strains are beginning to appear faster than new countermeasures can be deployed. Resistant strains of the western corn rootworm—once a minor pest in two states—are advancing 150 miles per year, immune to any currently available pesticide.[33] Resistant strains of the budworm have decimated cotton crops in Texas; the 1970 yield, for example, was the lowest since W.W. II. Overall, the use of pesticides on crops has increased 12 times since W.W. II, to more than 600 million pounds per year; but for some crops (e.g., corn), the losses from insects have more than tripled.[34] In 1976, the National Academy of Sciences released a five-volume report warning of imminent technological breakdown in the chemical control of pests.[35] It appears that pesticide technology is being "depleted" as it encounters the limits set ultimately by nature.

The Green Revolution

The development of the supercrops, together with the attendant cultivating technology, has been termed the Green Revolution.[36] Given the high productivity of industrial agriculture, it is perhaps not surprising that the Green Revolution has been urged on the rest of the world. Indeed, the enthusiasm has occasionally bordered on euphoria:

> Over the past 200 years the U.S. has had the best, most logical and the most successful program of agricultural development anywhere in the world. Other countries would do well to copy it.[37]

It is understandable that a hungry world has responded to such prompting. The recent history of world agriculture is dominated by a single theme: the exportation of the Green Revolution to all corners of the globe. India, to cite but one example, adopted supercrops in the mid-1960s. In the short span of a decade, supercrops accounted for more than 75% of Indian wheat production.[38] Accordingly, fertilizer consumption in India increased nearly 10-fold in the same period.[39] A similar pattern is repeated worldwide.[40] But as we will develop more fully in the next chapter, to the degree that exportation of the Green Revolution has succeeded, nonindustrial nations now also depend on fossil fuels, chemicals, and pesticides for food.

Energy Equals Food

Nearly 25% of global energy expenditure is currently directed toward food production.[41] In nonindustrial nations, where overall energy conversion is less, nearly 60% of the total energy budget is spent on food,[42] and in rural India the figure may approach 80%.[43] But traditional agricultures rely on sustainable energy sources such as animal power, manure, and human labor. In contrast, the industrial nations now depend almost entirely on fossil fuels for food, and they may therefore absorb the uncushioned impact of the coming energy scarcity.

One way to gauge the magnitude of the coming change is to analyze the U.S. food system in terms of the energy it uses. It is not an easy task, for as our own brains depend on our bodies for existence, our agriculture is inseparably integrated into the infrastructure of industrialism. We could compute the calories used only by the brain, but the figure is hardly meaningful unless the rest of the body is alive and well. Likewise, industrial agriculture cannot realistically be viewed in isolation from the rest of industrial society; its status is too heavily influenced by the health of the whole system. We can, however, begin to appreciate the immediate energy cost of our food by reducing the U.S. food system to its essential component steps (Figure 9.8) and assessing the energy used by each Table 9.1). Several recent analyses are available to assist in such accounting.[44]

Food Growing

The energy consumed in actually growing food on farms is a surprisingly small fraction of the total energy spent on food. It is estimated variously as from 0.5%[45] to 3.1%[46] of the annual U.S. energy use. Approximately three-quarters of U.S. farm energy is expended in the form of fuel, electricity, and fertilizer.[47]

Food Processing

In traditional agricultures, food is eaten near where it is grown and hence little energy need be expended following the harvest. Industrial societies, in contrast, must move food from farms to cities before it is marketed and consumed. Most foods require some kind of processing to withstand the journey; accordingly, in parallel with industrial farming, a food-processing industry has grown up in the U.S. This industry is now the fourth-largest industrial energy user in the U.S., after metals, chemicals, and oil. It consumes an estimated 3.1%[48] to 5.8%[49] of the total U.S. energy budget.

To assess the impact of the food-processing industry on our lives, we need only walk the aisles of any modern U.S. supermarket. Shelves and bins are packed with row upon row of processed foods: canned, bottled, pickled, refined, bagged, boxed, and frozen. Because of fossil fuels, most foods sold by supermarkets are processed, and the American diet has followed suit. But the food-processing industry has been shaped by the same economic force that has molded our farming enterprise. Accordingly, the laudable goal of simply protecting food from deterioration during transport has long since ceased to guide the industry. The goal now is to maximize profit.

In his book *Food for Nought*, Hall documents the consequences of modern food processing for human nutrition.[50] Food preservatives, for example, are added to processed foods not to protect the consumer, but to reduce the need for careful sanitation during processing and to increase shelf life. With manufacturing and marketing costs thus reduced, profits are increased, but the effects on human health are ominous. White bread is stripped of its nutritive value, bleached with poisonous chloroform gas, and filled with chemicals to facilitate manufacturing. Synthetic foods, sometimes literally infiltrated with plastic, are peddled to an uninformed public.[51] Livestock are packed into feedlots to maximize production efficiencies and then treated with antibiotics in order to combat infectious diseases that thrive under such crowded and unsanitary conditions. When we eat commercially produced meat, we ingest these same antibiotics. The continuous presence of these antibiotics in our blood destroys susceptible bacteria, but also selects for resistant strains of disease organisms, much as pesticides select for resistant strains of plant pathogens. Antibiotics thus lose their power to combat disease organisms in life-threatening situations. The U.S. Food and Drug Administration has moved to ban the use of antibiotics in livestock, but a regulatory agency with a budget measured in millions cannot effectively monitor a multibillion-dollar food industry. Thus most of the 1,800 chemicals that are routinely added to American food during processing have not been evaluated for their effects on humans.[52] The food-processing industry may be seen as a natural outgrowth of an industrialized agriculture, but it may be inadvertently depleting our most precious natural resource—human health.

Food Preparation and Transportation

Once food is processed and marketed it is prepared in homes and institutions, at an annual energy cost of 3.7%[53] to 4.7%[54] of the U.S. energy budget. But the energy used for preparing food pales before that needed for transportation. We have seen how the railroads transformed American farming early in our history. Transportation remains the basis of our industrial agriculture, by providing the essential link between each step in the food production cycle (Figure 9.8). Transportation of food is estimated conservatively to consume 0.4%[55] to 2%[56] of the annual U.S. energy budget, but these figures omit the transport of essential farm inputs and the use of private automobiles for food shopping, which are difficult to assess. To approach the matter somewhat differently, it is estimated that half the U.S. truck fleet is occupied with the transport of food or food-related items.[57] As we saw in Chapter 3, trucks use 21% of the transportation energy budget, which is in turn about one-fourth of the total. Thus 2.6% of the total energy budget goes toward truck transportation of food. If we include the energy that propels automobiles to and from the supermarket, food-related transportation is probably double the above estimate, i.e., 5.2%. If we also include the energy and resources used to build the trucks and cars, then the transportation of food becomes one of the largest single energy users in the economy. As we saw also in Chapter 3, transportation is 96% dependent on the liquid fuels that are in shortest supply. Thus food transportation is not only a heavy user of energy, it is also among the most vulnerable to oil depletion.

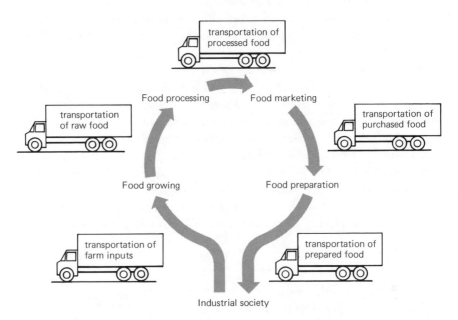

9.8   The food production cycle of an industrial nation such as the United States. Note how each step is linked to all others by the critical function of transportation.

Table 9.1 summarizes this accounting of energy consumption by each sector of the food industry. As shown there, the U.S. food system consumes from 7.7% to 15.5% of the total annual energy budget. The actual value probably falls between these extremes, and is estimated variously at 12.4% (1963 data)[58] and 12.8% (1970 data)[59] of the total. In terms of aggregate energy expenditure, the food industry is thus the largest in the U.S.

Table 9.1. ENERGY USE IN THE U.S.
FOOD SYSTEM [a]

Sector	ESTIMATED PERCENT OF ANNUAL U.S. ENERGY BUDGET	
	Low estimate	High estimate
Growing	0.5	3.1
Processing	3.1	5.8
Preparation	3.7	4.7
Transportation	0.4	1.9
Totals	7.7	15.5

[a] Data calculated from J. S. Steinhart and C. E. Steinhart, "Energy Use in the U.S. Food System," *Science* 184 (1974): 307–16; and E. Hirst, "Food-related Energy Requirements," *Science* 184 (1974): 134–38.

Diminishing Returns

In this chapter we have focused on agriculture in the United States, but the trends that we have identified—increasing reliance on energy-intensive, high-technology methods, the consequent decline in farm population, the increase in farm size and corporate farming, and the rise of food processing—all appear to be general throughout the world. Similar trends are seen, with minor variations, in the history of industrialized nations such as Japan[60] and England.[61] Likewise, similar trends are well developed in the USSR,[62] and are beginning to unfold in China[63] and other developing nations.[64]

But as energy and technology have come to shape world agriculture, we have begun to learn their limits. Pesticides furnish a classic example of how technology can be considered a resource that is subject to the same laws of depletion discussed earlier (see Chapter 8). Likewise, when increasing quantities of fertilizer are applied to crops, yields increase more slowly and eventually actually decrease with greater fertilizer applications.[65] As a result of such diminishing returns, absolute productivity on experimental farms has been static for a decade,[66] and the energy *efficiency* of the U.S. food system—i.e., the ratio of output to input energy—has declined steadily since the beginning of the energy era. In 1940 five calories of fossil fuel energy were invested for every calorie returned in food, but by 1970 the ratio had almost doubled, to nine to one

(Figure 9.9). As a consequence, farm output has leveled off despite increasing energy input (Figure 9.10). As for the future,

> Further applications of energy in the U.S. food system are likely to yield little or no increase in the level of productivity.[67]

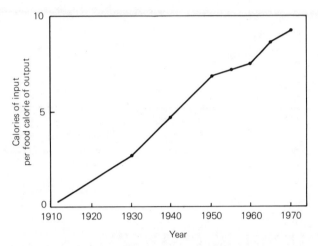

9.9 Energy "subsidy" to the U.S. food system needed to obtain one calorie of food. The subsidy has risen steadily in this century, implying a steady reduction of energy efficiency. (from J. S. Steinhart and C. E. Steinhart, "Energy Use in the U.S. Food System," *Science* 184 [1974]: 307–16, Figure 4)

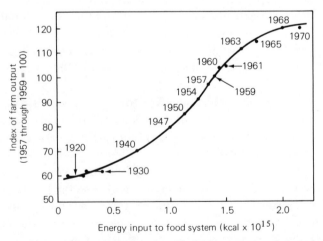

9.10 Farm output versus energy input in the U.S. food system over the past half century. Note how the curve reaches a plateau, implying a condition of diminishing returns. (from J. S. Steinhart and C. E. Steinhart, "Energy Use in the U.S. Food System," *Science* 184 [1974]: 307–16, Figure 2)

The problem of reaping less food from more energy is somewhat academic; industrial civilization may soon be confronted with the need for more food from less energy. For as we have seen, we are exhausting the fossil fuels that have in the past formed the basis of our food-producing system. What will become of our food supply, and the industrial civilization on which it depends, as energy and resources become depleted and therefore more expensive in the coming decades? It is to this critical question that we now turn.

NOTES

1. J. D. Chambers and G. E. Mingay, *The Agricultural Revolution, 1750–1880* (New York: Schocken Books, 1966).
2. Ibid.
3. L. Tweeten, *Foundations of Farm Policy* (Lincoln, Neb.: University of Nebraska Press, 1970).
4. Ibid.
5. B. F. Johnson and P. Kilby, *Agriculture and Structural Transformation* (New York: Oxford University Press, 1975).
6. Tweeten, *Foundations of Farm Policy,* p. 128.
7. Ibid.
8. C. J. Pratt, "Chemical Fertilizers," *Scientific American* 212 (1965): 2–12; W. J. Chancellor and J. R. Gross, "Balancing Energy and Food Production, 1975–2000," *Science* 192 (1976): 213–18.
9. S. H. Wittwer, "Food Production: Technology and the Resource Base," in *Food: Politics, Economics, Nutrition and Research*, edited by P. H. Abelson (Washington, D.C.: American Association for the Advancement of Science, 1975), pp. 85–90; D. Pimentel, L. E. Hurd, A. C. Bellotti, M. J. Forster, I. N. Oka, O. D. Sholes, and R. J. Whitman, "Food Production and the Energy Crisis," in *Food: Politics, Economics, Nutrition and Research*, edited by P. H. Abelson (Washington, D.C.: American Association for the Advancement of Science, 1975), pp. 121–27; R. W. F. Hardy and U. D. Havelka, "Nitrogen Fixation Research: A Key to World Food?" in *Food: Politics, Economics, Nutrition and Research*, edited by P. H. Abelson (Washington, D.C.: American Association for the Advancement of Science, 1975), pp. 178–88.
10. D. Hayes, *Rays of Hope* (New York: W. W. Norton, 1977), p. 93.
11. D. Zwerdling, "The Day of the Locust," *Mother Jones*, August 1977, pp. 35–38.
12. Ibid.
13. Ibid.
14. Ibid.
15. G. Leach, *Energy and Food Production* (Washington, D.C.: International Institute for Environment and Development, 1975).

188   The Staff of Life

16. Zwerding, "The Day of the Locust"; J. Walsh, "U.S. Agribusiness and Agricultural Trends," *Science* 188 (1975): 531–34.
17. From "Country Trucks," a poem by Monica Shannon, in P. Roberts *Roberts's English Series,* Book C (Sacramento, Calif.: State Department of Education, 1967), p. 14.
18. Tweeten, *Foundations of Farm Policy,* p. 238.
19. R. C. Fellmeth, ed., *Power and Land in California* (Washington, D.C.: Center for the Study of Responsible Law, 1971).
20. Ibid.
21. Tweeten, *Foundations for Farm Policy.*
22. J. S. Steinhart and C. E. Steinhart, "Energy Use in the U.S. Food System," *Science* 184 (1974): 307–16.
23. Zwerdling, "The Day of the Locust."
24. E.g., J. R. Harlan, "Our Vanishing Genetic Resources," in *Food: Politics, Economics, Nutrition and Research,* edited by P. H. Abelson (Washington, D.C.: American Association for the Advancement of Science, 1975), pp. 157–60.
25. E.g., P. R. Day, ed., *The Genetic Basis of Epidemics in Agriculture* (New York: New York Academy of Science, 1977).
26. D. R. Marshall, "The Advantages and Hazards of Genetic Homogeneity," in ibid., pp. 1–20.
27. S. Beart, "Genetic Peril in Cattle," *Environment* 19 (1977): 2–3.
28. Zwerdling, "The Day of the Locust."
29. Ibid.
30. E.g., E. J. Kormondy, *Concepts of Biology,* 2nd ed. (Englewood Cliffs, N.J.: Prentice-Hall, 1976).
31. D. N. Duvisk, "Major United States Crops in 1976," in *The Genetic Basis of Epidemics in Agriculture,* edited by P. R. Day (New York: New York Academy of Science, 1977), pp. 86–96.
32. Day, *The Genetic Basic of Epidemics in Agriculture.*
33. Zwerdling, "The Day of the Locust."
34. Ibid.
35. L. J. Carter, "Pest Control: NAS Panel Warns of Possible Technological Breakdown," *Science* 191 (1976): 836–37; H. M. Schmeck, Jr., "Insects' Resistance to Pesticides Held Rising Alarmingly," *New York Times,* 6 February 1976, p. 1.
36. P. R. Jennings, "The Amplification of Agricultural Production," in *Food and Agriculture* (San Francisco: W. H. Freeman, 1976), pp. 125–33; D. J. Greenland, "Bringing the Green Revolution to the Shifting Cultivator," *Science* 190 (1975): 841–44; P. Jennings, "Rice Breeding and World Food Production," in *Food: Politics, Economics, Nutrition and Research,* edited by P. H. Abelson (Washington, D.C.: American Association for the Advancement of Science, 1975), pp. 95–98; N. L. Brown and E. R. Pariser, "Food Science in Developing Countries," in *Food: Politics, Economics, Nutrition and Research,* edited

by P. H. Abelson (Washington, D.C.: American Association for the Advancement of Science, 1975), pp. 93–103.

37. E. O. Heady, "The Agriculture of the U.S.," in *Food and Agriculture* (San Francisco: W. H. Freeman, 1976), p. 97.

38. J. D. Gavin and J. A. Dixon, "India: A Perspective on the Food Situation," in *Food: Politics, Economics, Nutrition and Research*, edited by P. H. Abelson (Washington, D.C.: American Association for the Advancement of Science, 1975), pp. 49–57.

39. Ibid.

40. Day, *The Genetic Basis of Epidemics in Agriculture*.

41. D. Pimentel, E. C. Terhune, R. Dyson-Hudson, S. Rochereau, R. Samis, E. A. Smith, D. Denman, D. Reifschneider, and M. Shepard, "Land Degradation: Effects on Food and Energy Resources," *Science* 194 (1976): 149–55.

42. A. Makhijani, *Energy and Agriculture in the Third World* (Cambridge, Mass.: Ballinger, 1975).

43. R. Revelle, "Energy Use in Rural India," *Science* 192 (1976): 969–75.

44. See above, notes 8, 9, 15, and 22; D. Pimentel, W. Dritschilo, J. Krummel, and J. Kutzman, "Energy and Land Constraints in Food Protein Production," *Science* 190 (1975): 754–60; A. J. Fritsch, L. W. Dujack, and D. A. Jimerson, *Energy and Food* (Washington, D.C.: Center for Science in the Public Interest, 1975); E. Hirst, "Food-Related Energy Requirements," *Science* 184 (1974): 134–38.

45. Hirst, "Food-Related Energy Requirements."

46. Steinhart and Steinhart, "Energy Use in the U.S. Food System."

47. Ibid.

48. Hirst, "Food-Related Energy Requirements."

49. Steinhart and Steinhart, "Energy Use in the U.S. Food System."

50. R. H. Hall, *Food for Nought* (Hagerstown, Md.: Harper and Row, 1974).

51. Id., "The Food Fabricators. II," *Environment* 18 (1976): 17–36.

52. Hall, *Food for Nought*.

53. Hirst, "Food-Related Energy Requirements."

54. Steinhart and Steinhart, "Energy Use in the U.S. Food System."

55. Fritsch et al., *Energy and Food*.

56. Steinhart and Steinhart, "Energy Use in the U.S. Food System."

57. Ibid.

58. Hirst, "Food-Related Energy Requirements."

59. Steinhart and Steinhart, "Energy Use in the U.S. Food System."

60. K. Ohkawa, "Phases of Agricultural Development and Economic Grownth," in *Agriculture and Economic Growth*, edited by K. Ohkawa, B. F. Johnston, and H. Kaneda (Tokyo and Princeton, N.J.: University of Tokyo and Princeton University Presses, 1970).

61. Leach, *Energy and Food Production*.

62. D. K. Shipler, "Soviet Decrees More Emphasis on Specialized Modern Farms," *New York Times*, 2 June 1976, p. 6; id., "Soviet 'Factories' Press the

Industrialization of Agriculture," *New York Times*, 12 October 1976, p. 58.

63. F. Butterfield, "China in Major Drive to Mechanize Farms," *New York Times*, 25 December 1976, p. 1.

64. J. Kandell, "Brazil's Agriculture Expands Fast, But Mostly for Benefit of Well-to .do," *New York Times*, 16 August 1976, p. 2.

65. See, e.g., above notes 9, 22, and 36.

66. V. K. McElheny, "Scientific Dead-end Faces U.S. in Its Agricultural Technology," *New York Times*, 3 May 1976, p. 1.

67. Steinhart and Steinhart, "Energy Use in the U.S. Food System," p. 312.

Chapter 10
THE SEEDS
OF CHANGE

As in the past, civilizations of the future will revolve around their systems of food production. To assess the future of industrial civilization it is helpful to reduce food production to the form of a simplified equation, with inputs on the left and the product, food, on the right. In its most elementary form, the equation contains but one input term, the primary natural resources (Figure 10.1). In Chapter 6 we saw that these resources consist of the basic elements—climate, land, water, air, sunlight, and the oceans—and in Chapter 2 we saw how nature converts these inputs to food via the food cycle. We have also seen how the human species has accelerated the food cycle in two historic steps. First, when humanity turned to agriculture to feed its growing numbers, human labor was increased as an input in the food equation (Figure 10.1). Human labor was of course an essential component of food production even in "primitive" hunter/gatherer socieities. But as noted in Chapter 2, hunter/gatherers could meet their food needs with a minimum of labor—two or three days per week. Second, the Industrial Revolution introduced new inputs, namely, capital and energy (Figure 10.1). Capital takes the form of tractors, trucks, and food-processing factories, all fashioned ultimately from natural resources. Energy is represented indirectly in the production of capital and agricultural technology, and directly by the fossil fuels that power our farm machines and make our fertilizers. As documented in the preceding chapter, the heavy reliance on capital and energy has reduced the need for human agricultural labor to a minimum.

Type of society	Nature of food-producing system
Hunter/gatherer	primary natural resources $\longrightarrow$ food a
Preindustrial agriculture	primary natural resources $+$ human labor $\longrightarrow$ food b
Industrial agriculture	primary natural resources $+$ human labor $+$ capital and energy $\longrightarrow$ food c

10.1 Evolution of human food-production systems. Hunter / gatherers depended on the primary natural resources and a minimum of human labor (*a*). Subsequent changes entailed the addition of more intensive human labor in preindustrial agriculture (*b*), and the addition of capital and energy (and the consequent reduction of human labor) in industrial agriculture (*c*).

The Agricultural Equation

The agricultural equation we have just formulated (Figure 10.1c) permits a deduction that is central to the thesis of this book. As is true of all such equations, when any input term on the left is diminished, the output on the right must also diminish unless some other input term can be increased to compensate. In other words, if capital and energy are withdrawn from agriculture, then either less food will be produced or some other input to agriculture must increase. Besides capital and energy, only two inputs remain in the agricultural equation: primary natural resources and human labor. We have already examined the primary natural resources in some detail in Chapter 6. Global climate, which has been optimal for agricultural productivity in the 20th century, is expected by climatologists to become more variable over the next few decades. Variable weather in turn can[1] and lately has[2] played havoc with agricultural productivity. Industrial nations already cultivate most of their arable lands, and the capital and energy necessary to bring new land into cultivation are increasingly scarce (Chapter 6). The quality of our water and air cannot be expected to improve rapidly, as we have discussed (Chapter 6), and the tiny fraction of the sun's energy that is intercepted by our planet cannot change. In short, we cannot reasonably expect increases in the primary resource base to compensate for the withdrawal of capital and energy from the agricultural equation.

The only remaining variable in the agricultural equation is human labor. In the nonindustrial world, human labor already dominates food production, with fully 80% of the populace typically directly engaged in agriculture.[3] But in the industrial nations we have seen how fossil fuels have steadily displaced human labor (Chapter 9), with the result that the few grow food for the many. The depletion of energy and resources would seem to imply the reversal of the two-century trend of substituting energy for farm labor. As energy and resources become increasingly scarce, industrial societies may have to divert labor from industry to agriculture if they are to avoid a precipitous drop in food production. Thomas Jefferson's vision of a nation of farmers laboring in the earth[4] may be realized, although for reasons that Mr. Jefferson could not have forseen.

The Decentralization of Agriculture

The redirection of human labor toward agriculture is a straightforward, logical consequence of energy and resource depletion, but it is only one step in an interrelated sequence that may transform our civilization. We have seen, for example, how the food-production cycle of an industrial society relies heavily on mechanized transportation (Chapter 9, Figure 9.8). Such transportation in turn depends almost exclusively on the liquid fuels that are in shortest supply (Chapter 3). As fossil fuels and metals become depleted, their cost may be expected to soar (Chapter 8). If so, then it will become exponentially more expensive to build and operate the trucks that carry our food from the site of production to the site of consumption. The cost of transportation will thus comprise an increasing fraction of the food bill, reducing the distance over which food can be moved economically. Winter tomatoes from Mexico, citrus fruits from Florida and Texas, lettuce from California, Canadian bacon, and Australian beef—all may become memories of a more affluent past. The depletion of energy and resources implies not only that more people will labor in the fields, but it also implies that food will be consumed progressively closer to where it is grown.

In the early stages of transition backyard gardens may bloom again, furnishing a suitable if transient means of food self-sufficiency. People may depend increasingly on food that they themselves grow. But as we shall develop in the remainder of this book, the decline of energy and resources may compel the emergence of a new and much different social order, in which the lives of most people revolve around food production. In the absence of transportation, the food system would of necessity decentralize. Food production and consumption would have to occur in the same physical locality: each region, state, and eventually county would be required by the lack of cheap transportation to supply most of its own food needs; and the consequent increase in regional independence might exert unrelenting pressure for the decentralization of all institutions in our industrial civilization.

The Decline of Food Processing

We have examined how the processing of foods is a natural outgrowth of the need to transport food over long distances and store foods for long periods during marketing (Chapter 9). The decline of transportation implies a corresponding decline in the need to process food. We have also seen that the food-processing industry is now the fourth largest in the U.S., and thus requires a prodigous flow of energy. As energy becomes more expensive in the coming decades, the cost of food processing may be expected to increase. The depletion of energy and resources may thus render the food-processing industry both unnecessary and uneconomic. The industry may shrink accordingly; Americans may therefore of necessity eat fewer processed and refined foods, and their health may benefit correspondingly.

Changing What We Eat

Perhaps the most immediate impact of energy and resource depletion will be manifest in our diet. As we saw in Chapter 2, our food is derived from the food chain, in which solar energy is "upgraded" first to plants and then to animals. At each step, or trophic level, of the food chain, the laws of thermodynamics exact their toll of "lost" energy, dissipated in the form of unusable heat. These losses are substantial: 1,000 calories of sunlight yield but 10 or 20 plant calories, which in turn yield but 1 or 2 calories of energy stored in animal flesh. In other words, foods that occupy the highest trophic levels in the food chain require by far the most energy to produce.

Before the advent of fossil fuels, diets were necessarily drawn from the lower end of the food chain. In the U.S. in 1876, 55% of all dietary calories were derived from carbohydrates, or "starches."[5] Our great-grandparents existed on a diet composed largely of bread, potatoes, fruits, and vegetables. But as fossil fuels have increasingly come to dominate our lifestyle, our diets have shifted toward energetically expensive foods. In the U.S. and other industrial nations today, only 20% of all food calories are consumed in the form of carbohydrates.[6] The remainder of the American diet is composed largely of meat.

The feedlot beef that is considered a dietary staple in the U.S. today is among the most energetically expensive foods ever routinely eaten by the human species. Feedlot beef requires some 12 input calories for each food calorie returned (Figure 10.2). In other words, such livestock consume far more food energy than they yield. Similarly, livestock consume more protein than they return: 14–21 pounds of edible plant protein are

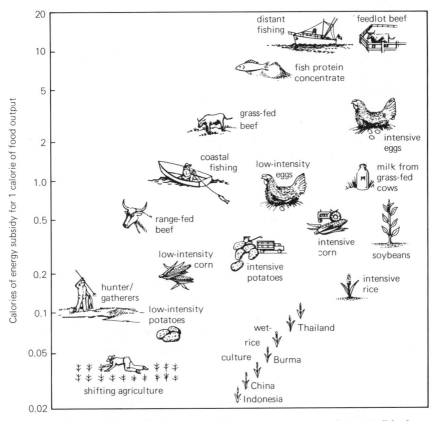

10.2 Energy costs of various food crops. Note that in energetic terms, "primitive" food systems (lower left) are more efficient in generating food calories than "modern" food systems (upper right). (from J. S. Steinhart and C. E. Steinhart, "Energy Use in the U.S. Food System," *Science* 184 [1974]: 307–316, Figure 5)

required to produce a single pound of animal protein.[7] And yet, because
fossil fuels have been abundant and cheap, 95% of the U.S. grain crop is
fed to livestock,[8] and the biomass of beef in the U.S. is more than double
the biomass of the human population (calculated from U.S. cattle population reported by Robbins).[9]

The meats, poultry, fish, and dairy products that occupy the top of
the food chain require more energy and protein input than they deliver in
return (Figure 10.2). Therefore, these energy-expensive products do not
yield net food. But because fossil fuels have been abundant and therefore
cheap, industrial civilization has been temporarily exempted from the historical necessity of producing net food. As fossil fuels have generated capital (Chapter 5), industrial nations have spent it on meat; there is a nearly
perfect linear relationship between per capita national income and per
capita national meat consumption (Figure 10.3).

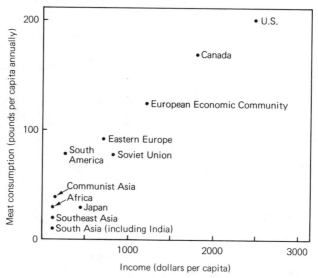

10.3 Meat consumption rises with increasing national wealth. (from U.S. Dept. of Agriculture data)

As energy and natural resources decline in availability, subsidization
of food production by fossil fuels may no longer be economically feasible.
Humanity may therefore again be required to generate net food energy as
the price of survival. We may have to rely on energetically less expensive
food sources, such as the root and grain crops that occupy the lowest
trophic levels of the food chain. These crops are exceptionally efficient
energy converters. Corn, soybeans, rice, and potatoes all require a fraction of a calorie of input energy for each food calorie returned (Figure
10.2). Certain root and grain crops are also efficient producers of precious

protein. Soybeans, for example, the staple of many Asian diets for centuries, generate 356 pounds of edible protein per acre of land per year, compared with 20 pounds of protein from beef (Figure 10.4). As fossil fuels are depleted, we may no longer be able to afford the energetic extravagance of drawing our diet from the top of the food chain. We may of economic necessity return to a starchier but nutritionally adequate diet that is based largely on the root and grain crops.

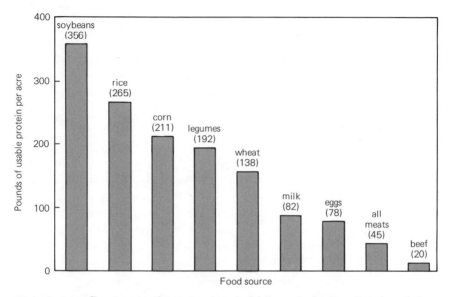

10.4 Protein efficiencies (productivity per acre) of different foods. Note the relatively low efficiency of meat in comparison with plants. (calculated from U.S.D.A. per-acre yield statistics, 1971–1974 [Washington, D.C.: U.S. Government Printing Office], and WHO/FAO/UNICEF Protein Advisory Group Bulletin #6, 1971)

Enriching the Earth

Regardless of the energetic content of a given food source, food production cannot be maintained unless the nutrients in the soil that are incorporated into food are regularly replenished. One ton of wheat grain takes from the soil the equivalent of 40 pounds of nitrogen; a ton of cattle, 54 pounds. These losses must be compensated for by regular additions of fertilizer. Industrial agriculture relies on synthetic, or chemical, fertilizers, the most significant of which is nitrogen.[10] Nitrogen fertilizer is currently manufactured from synthetic ammonia, which is in turn made from gaseous nitrogen and hydrogen. Nitrogen gas is plentiful, comprising some three-quarters of the earth's atmosphere; but hydrogen can be economically obtained in quantity from only one source at present—fossil fuels. Most nitrogen fertilizer now used by industrial agriculture is made from the fossil fuel that is in shortest supply, natural gas. The manufacture of nitrogenous fertilizers currently accounts for 2% of U.S. natural-gas consumption, and 5% of worldwide consumption.[11] Coal and petroleum can also be used to generate hydrogen, but the process is technically awkward and substantially more expensive.[12]

As fossil fuels become depleted, it may no longer be economic to manufacture nitrogenous fertilizers as we have in the past. From 1971 to 1974 alone, fertilizer prices increased eightfold,[13] driven by a sixfold increase in oil prices during the same period. As energy is depleted, the price of fertilizers manufactured from fossil fuels may be expected to continue to increase. If food production is to be maintained, we may have to look elsewhere for essential fertilizers.

By far the cheapest alternatives to fossil-fuel fertilizers are animal manures, green manures, and human feces ("night soil").* Green manures, such as legumes (peas) and alfalfa, are cover crops that are especially good at converting atmospheric nitrogen into a form the soil can use. But to use green manures requires the temporary withdrawal of land from food production. Animal manures are available conveniently and in volume, but they also require that land be removed from the production of human food in order to grow animal fodder. On an ecologically balanced, totally self-sufficient farm, approximately one-fourth the arable acreage is required for growing animal fodder.

We see that the two major fertilizer alternatives to fossil fuels require a substantial reduction of land for growing human food. This reduction can be compensated for only by using existing land more intensively, by intercropping and double-cropping. These farming practices, employed widely in nonindustrial nations like China, are labor intensive rather than energy intensive. They may have the effect of accelerating the diversion of labor from industrial pursuits to agriculture.

The Despecialization of Agriculture

The U.S. currently has a copious source of animal manure generated by the livestock industry. Pimentel and associates report that nearly two

---

* Safe, odorless, nonflush toilets that convert human excrement into a rich, powdery, dry fertilizer are already marketed under the names Clivus Multrum and Ecolet.

billion tons of livestock manure are produced each year in the U.S.[14] If half this manure could be returned to the earth it would more than satisfy the current fertilizer needs of the most important U.S. grain crop, corn. The major current costs of using animal manure for fertilizer—excluding the energy cost of the animal food—are associated with hauling and spreading. The energy cost of hauling and spreading manure within a 0.5- to 1.0-mile radius is calculated by Pimentel et al. as 400,000 kilocalories (kcal.) per acre per year, as compared with a million kcal. to manufacture and spread a comparable quantity of chemical fertilizer.[15] For distances over a few miles, however, the energy costs of transporting manure begin to exceed energy costs of chemical fertilizer. As energy and resources become depleted, however, their increased cost may discourage the use of chemical fertilizer; and higher transportation costs may require that animal manure be generated near where it is used—that is, on the same farms.

The depletion of fossil fuels thus implies the need for ecologically balanced farms. In the absence of energy-intensive pesticides farmers may have to intermix several crops to reduce damage from insects and other pathogens. In absence of chemical fertilizers farmers may have to grow their own, in the form of animal manures. The dependence of people on locally grown food, also implied by the decline of transportation, will presumably also influence farms to diversity. As the specialization of the American farm was made possible by the rise of transportation, despecialization is the logical outcome of the decline in transportation that is implied by the depletion of energy and resources.

Demechanization and the Rise of Animal Power

Mechanization of the U.S. farm is a recent chapter in American history. As we saw earlier (Chapter 9), tractors and trucks began to replace horses and mules well into the 20th century, within the memories of many people now alive. But as fossil fuels and natural resources made mechanization possible, so the depletion of energy and resources may render mechanization increasingly uneconomic. Pimentel and colleagues have documented how mechanization of agriculture requires vastly increased energy inputs.[16] For example, applying a chemical to an acre of corn requires 18,000 kcal. using a tractor and sprayer, but less than 300 kcal. if applied by a hand sprayer.[17] Although nonmechanized labor in this example uses but 1.7% the energy of mechanized methods, it takes much longer. At current energy and labor prices, the cost of hand application is four times that of tractor application.[18] As the price of energy goes up, however, human labor will become steadily more economic.

Likewise, the use of animals in place of machines may gradually return to American agriculture as the relative cost of mechanization increases. Animal power is not without costs, requiring a substantial expenditure of energy in the form of animal food.[19] It is the *relative* price, however, that dictates which is more economic, and as energy prices soar the relative price of animal food (solar energy) will decline compared with tractor fuel (fossil fuel energy). Moreover, as metals are exhausted (Chapter 7) the capital costs of tractors may become prohibitive, especially in comparison with the cost of a horse or mule. Already a new, multipurpose tractor can cost $50,000–$100,000. The need to use animal manures for fertilizers may return animals to the American farm. As the price of energy and resources soars in the coming decades and mechanization becomes correspondingly more costly, the same animals that provide essential fertilizer may also draw our plows and wagons once again.

Economics:
The Engine of Change

In this and preceding chapters I have repeatedly suggested that the exhaustion of energy and resources is not necessary to the transformation of industrial civilization, and that, instead, the soaring prices symptomatic of depletion provide a sufficient driving force. The arguments developed above suggest how this could come to pass. When the price of energy and resources increases, then the relative price of everything that depends on energy and resources also increases. Therefore, the relative price of other things—notably human labor—declines. Thus it becomes more "economic," that is, less costly, to use labor-intensive rather than energy-intensive methods. Tasks performed with human labor take longer to perform than tasks performed using fossil-fuel energy, since the latter is more "concentrated" energy. Therefore, accomplishing the same job requires the participation of more people—and hence the diversion of human labor into agriculture.

Food Prices: A Scenario

But agriculture is not the only enterprise that would be affected by the depletion of energy and resources. Instead we may expect that the entire price structure of our economy—i.e., the relative prices of all different goods and services—will shift to reflect the new realities. We have seen, for example, how a sixfold increase in the price of energy prompted an eightfold increase in fertilizer price in the early 1970s.[20] Because our food-production system is so thoroughly dependent on energy (Chapter 9), the price of food is likewise linked tightly to the price of energy.

The exact relation between food cost and energy cost is uncertain and deserves detailed study. But preliminary indicators suggest that when energy prices increase, food prices go up half again as much.[21] We can use this relationship to estimate the future cost of food if we know the future cost of energy. As we saw in Chapter 8, the future price of energy is uncertain; estimates range from a doubling or tripling in price by the year 2000, to a 100-fold increase. In the past generation, overall energy prices have increased 11-fold (Chapter 8). For the future let us assume a conservative fivefold energy price increase in the next two decades—a rate of increase of less than half that of the recent past. If energy prices increase five times, then according to the above relationship food prices will increase 7.5 times.

It is instructive to explore the consequences of a 7.5-fold food price increase for the average American family. In 1973, the intermediate budget for a family of four people included $2,532 for the food bill, 23% of its median income of $10,971.[22] If food costs 7.5 times as much in the year 2000, the same family will incur a food bill of $18,990. Between 1951 and 1971, median U.S. incomes increased linearly.[23] Linear extrapolation of the curve to the year 2000 yields a projected median income of $19,000, against a projected food bill of $18,990. In other words, after buying the groceries the average U.S. family of four would have exactly $10.00 with which to purchase the remaining necessities of life, including housing, clothing, medicine, transportation, education, and recreation.

"Differential Inflation" and the Price Structure

Long before this simplified scenario can unfold in its entirety, society will presumably have begun the adjustments required for its members to meet their basic survival needs, as we shall develop in the remainder of this book. But the price scenario presented above nevertheless illustrates a crucial generality. As the price of energy and natural resources is driven steadily upward by depletion, the prices of energy-intensive goods are expected to increase more rapidly than incomes. Therefore the depletion of energy and natural resources is not necessary to effect the transformation of our civilization: the relative price increase of energy and resources that may be expected to result from increasing scarcity is a sufficient driving force. Such a relative price increase would be expected to manifest itself as inflation, whose roots we examined in Chapter 5. But inflation will not affect all goods equally; instead, the rate of inflation would be expected to be greatest for those goods that are most dependent on energy—e.g., food. It is such "differential inflation" that may eventually alter the price structure of industrial economies beyond recognition.

How will these economic consequences of energy and resource depletion affect the average citizen of industrial nations? If the recent past is a reasonable guide, then the purchasing power of the individual will decline steadily as prices rise faster than incomes. People may earn more money, but more money may buy fewer goods and services. Such as erosion of individual purchasing power has already afflicted the economies of all industrial nations since the quadrupling of oil prices in the early 1970s. As a result, the proportion of people below the "poverty line" has steadily increased,[24] and the material standard of living in the U.S. and other industrial nations has declined for the first time in recent history.[25] Inflation has increased month after month in the recent past, led by the rising cost of housing and food.[26] As a consequence, the cost of the average new home is now approaching $60,000, well beyond the economic reach of the average American.[27] When the average American can no longer afford to purchase sufficient food, it seems likely that a radical restructuring of the social order will be close at hand.

A Return to the Past?

In terms of the quantity of food produced per unit time, there is no question that the industrial agriculture of today is the most bountiful the world has ever known. But sheer quantity is a narrow measure of productivity. A much broader measure is energetic efficiency, defined as the energy returned per unit of energy invested. In terms of energetic efficiency, industrial agriculture is the least productive food system in human history. We have seen how the contemporary U.S. food system requires nearly 10 calories of input energy for every calorie returned in food. In striking contrast, hunter/gatherer cultures and shifting agricultural societies of the past were a hundred times more efficient, requiring less than one-tenth of a calorie of input energy for each food calorie produced (Figure 10.2).

What primitive societies gained in efficiency, they of course lost in total food output. One consequence was a much smaller global population than is supported by our planet today. This finite earth cannot sustain many such nonagricultural societies at the same time; there is not enough food-producing capacity. Thus, however desirable a return to the distant past might seem to some, it is physically impossible without a manyfold reduction of world population. In any case the romantic view of the simple, agrarian life of the past is effectively dispelled by the gripping accounts of preindustrial England, where life was characterized by high child mortality, ill health, periodic plague, frequent economic depression, oppression of one class by another, and unending physical hardship from birth to premature death.

The future civilization whose boundaries we have begun to sketch in this chapter bears a superficial resemblance to the past, if only because they share a profound commonality—reduced reliance on energy and natural resources. But here, I believe, the similarity may end. We are separated from the past by an unbreachable gulf that is traceable to entropy and represented by knowledge. In thermodynamic terms, the fossil fuels burned during the Industrial Age have been converted to people, machines, and ultimately knowledge. Knowledge is the most highly organized, concentrated, and precious form of energy; it occupies the highest "trophic level" in the material universe. The expenditure of energy during the Industrial Age has generated unprecedented knowledge of ourselves, the earth, and the universe. It seems likely that every aspect of life in the future will be unavoidably touched, colored, and shaped by the vast material knowledge generated during the Industrial Age.

Models from the Present

The need to sustain a gigantic human population, coupled with the existence of a vast material knowledge, suggests that our future cannot represent a return to the past. More useful models for the future are furnished by existing nonindustrial cultures. Again we see the surprising pattern that nonindustrial agricultures are more energy efficient than industrial agricultures, although they generally produce less food per unit land area (Figure 10.5). But among existing agricultural systems, the contemporary Chinese example stands apart from all others. Not only is its energetic efficiency—that is, the ratio of output energy to input energy—higher than any other on earth; in addition, its absolute productivity per unit land area exceeds even the most productive industrial agriculture (Figure 10.5). These astonishing efficiencies have been achieved by intensive

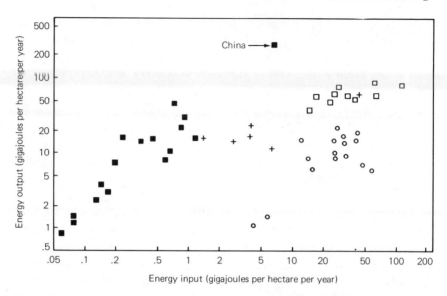

10.5   Energy input versus energy output for various industrial (□), British (○), semi-industrial (+) and nonindustrial (■) agricultural systems. Note the exceptional productivity of the Chinese system. Logarithmic scales. (from G. Leach, *Energy and Food Production* [Washington, D.C.: International Institute for Environment and Development, 1975])

application of human labor to agriculture, including such labor-intensive practices as intercropping and double-cropping, combined with intensive land use and irrigation.[28]

The People's Republic of China is not only the largest nation on earth, with a population approaching one billion; it is also one of the few nonindustrial nations that can feed itself. But the Chinese food system evolved in the context of a unique history, and its self-sufficiency has been bought only at a major cost in political regimentation and loss of the individuality and personal freedom that is so cherished in Western nations. Moreover, the Chinese apparently cannot resist the tempting fruit of industrialism. Mechanized agriculture and the use of fertilizers made from fossil fuels are on the rise, and the current Chinese leadership is apparently determined to industrialize its agriculture as rapidly as their potentially large domestic energy reserves will permit.[29] The same drive toward industrialism is seen the world over, as we shall discuss later. The Chinese example at least furnishes living evidence that a nonindustrial agriculture can support a large human population and an advanced human culture.

Bridges to the Future

If energy and resources are truly on the wane, then the choices that confront humankind are astonishingly direct. We can choose inaction, which is equivalent to burning our bridges before we reach them. Or we can build bridges to the future, bridges designed to bear the stress of what promises to be the most profound and rapid cultural transition in the human experience.

Food in Global Perspective

We can begin building bridges to an energy-scarce future by recognizing that history has cast the U.S. in the role of the world's breadbasket. Among industrial nations, only the U.S., Canada, and Australia export more food than they import.[30] The U.S. alone is responsible for 86% of this planet's surplus food production,[31] and North America accounts for 93% of the world's export of grain. The U.S. thus has a much broader monopoly on food than the Arab nations have on oil.

But while the U.S. is rich in food, we have seen how it depends increasingly on the fossil fuels and mineral resources supplied by other nations, most of which are nonindustrial. All the nations of the world are entangled in a network of interdependencies. Unless the U.S. continues to export food, the likelihood of global famine will increase; and in the event of global famine, the resulting social, economic, and political instability seems likely to interrupt the flow of imported energy and resources that powers our economy. If any semblance of world order is to be maintained during the approaching transition, the U.S. must continue to help feed the world's hungry, if only in the narrow pursuit of national self-interest.

Redefining the Green Revolution

The U.S. has responded to its historic role in part by developing and exporting an industrialized agricultural technology, as we have noted. Owing largely to this Green Revolution, food production has increased at similar rates in both industrial and nonindustrial nations for the past decade, although population growth has nullified the gain in nonindustrial nations.[32] But if the entire world converted to an industrial system of agriculture like our own, fully 80% of world energy conversion would be directed toward food production,[33] and the world's known reserves of petroleum would be exhausted in a decade.[34] There is little prospect of this

happening, for the Green Revolution has run aground on the high price of
energy. Owing to the increased cost of fossil fuels, fertilizer use in India
reached a plateau in 1972 and has remained static ever since.[35] The same
pattern is repeated on a global scale. Because of the increased cost of
energy, the nonindustrial nations spent nearly three times as much
money in 1974 for the same quantity of food and fertilizer that they imported in 1971 (Figure 10.6). Thus the Green Revolution, far from providing a panacea for hunger, has in reality created a cruel dilemma. As
energy prices spiral upward, the nonindustrial nations are increasingly
unable to purchase the energy-intensive farm inputs on which their food
supply now depends.

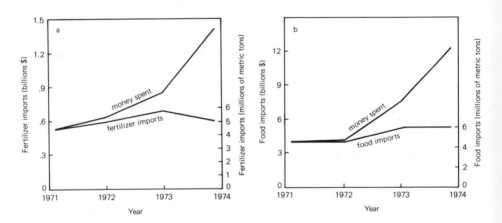

10.6   These graphs illustrate the impact of inflation on fertilizer (*a*) and food (*b*) imports by
developing nations. The money spent on each commodity has increased dramatically, while
the absolute imports have leveled off and even declined. (after W. D. Hopper, "The
Development of Agriculture in the Developing Countries," in *Food and Agriculture* [San
Francisco: W. H. Freeman, 1976], p. 137–144)

    The euphoria that initially accompanied the Green Revolution is
being replaced by an increasing awakening to its shortcomings. Even
when energy is inexpensive and readily available, an industrialized system of agriculture cannot function effectively when taken out of the context of an industrialized society. As we have seen (Chapter 9), industrial
agriculture depends on a food-processing industry, storage systems,
transportation networks, marketing mechanisms, a broad research base,
and a social organization and perspective that do not and cannot exist in a
nonindustrial society.[36] Now that energy is no longer inexpensive and

readily available, the Green Revolution is worse than futile—it is a tragic delusion. When the U.S. system of industrialized agriculture is transplanted to a non-industrial nation, the net effect is to increase dependence on an inaccessible resource (energy) and bypass an abundant one (labor), with consequent social dislocation. The Green Revolution has caused an increase in unemployment in nonindustrial nations as human labor is displaced by energy. It has caused an increase in the landless population as small farmers have been displaced by larger landowners. In Sonora, Mexico, for example, the Green Revolution has quintupled the average size of farms, to 2,000 acres.[37] And as we have noted, the Green Revolution has typically provided little relief from local hunger as additional food production is channeled into export markets by capital-starved nonindustrial economies. Throughout Central America and the Caribbean, the best half of the agricultural land now produces food for export to the U.S. and other industrial nations.[38] During the African Sahelian drought and accompanying famine of the early 1970s, hundreds of tons of fresh vegetables were exported from the area to the richer European market.[39]

If the Green Revolution is to help rather than hinder the approaching transition, it must be redefined and redirected to reduce rather than increase the dependency of world food production on energy. The goal must be to improve the efficiencies and productiveness of traditional agricultures, using knowledge and "soft" technologies that can be mastered and maintained by nonindustrial societies. If the transition is to be accomplished in an atmosphere of order, rather than social chaos, the fruits of the Green Revolution must reach those who need it most—the world's poor and hungry. Otherwise the Green Revolution is likely to cause far more human suffering and disorder than it has relieved.

Bringing
the Green Revolution Home

The agricultural systems of industrial nations are, as we have noted, thoroughly dependent on energy and resources. If energy and resources are truly on the wane, then a rational domestic farm policy must share with foreign policy the goal of reducing the dependence of agriculture on energy. Agricultural research in the U.S. is now focused on energy- and capital-intensive farming methods.[40] If it were refocused on labor-intensive, nonchemical farming, the research could benefit nonindustrial agricultures while simultaneously preparing for an energy-scarce future at home. U.S. agriculture now depends on the energy-intensive supercrops, as we have seen (Chapter 9). The land races that are so well suited to labor-intensive agriculture are becoming extinct. Industrial societies could maintain the precious genetic material on which plant breeding depends and at the same time prepare for the future by establishing worldwide seed banks to collect and preserve the land races. National seed banks already exist, but they are understaffed and inadequately funded.[41] Presently there is no organization in the world to husband the dwindling genetic resource of domestic animals.[42] The establishment and adequate staffing of such facilities could contribute importantly to the cause of human survival in an energy-scarce future.

If food production in the U.S. is to be maintained so that we can fulfill our historic role as the world's breadbasket while continuing to feed our own people, then the eventual diversion of domestic labor from industry to agriculture would seem unavoidable. Individuals can prepare for such a transition by educating themselves and others in the need; by acquiring the necessary tools and skills; and—where possible—by relocating on sufficient land. Schools and universities can prepare for such a transition by implementing appropriate educational programs and by redirecting agricultural research toward small-scale, nonchemical farming. Even in the context of industrial agriculture, experience has shown that "organic" farming can be as profitable as chemical farming.[43]

Land, Labor, and Tax Reform

Governments can prepare for the possible transition to a labor-intensive agriculture by designing and implementing land, labor, and taxation policies that are consistent with the requirements of an energy-scarce future. Tax policies, for example, could be shaped to favor the family farmer rather than the large corporate farm. Studies have shown that the small family farm is just as productive, acre for acre, as the large corporate farm.[44] Some states—notably North Dakota—have already banned corporate agribusiness outright.[45] The largest agricultural states could do well to follow this lead.

Some land policies favoring the establishment of small family farms already exist. For example, the Federal Reclamation Act of 1902 limits the size of farms irrigated by federally subsidized water projects to 160 acres. The law has never been enforced, however, with the result that the average size of federally irrigated farms in California exceeds 2,000 acres.[46] Hundreds of would-be family farmers stand ready to buy farmland if and when the law is enforced,[47] but the Carter administration shows strong signs of bowing to landed interests by opposing the law.[48] Unless the government can anticipate the demands of an energy-scarce future and implement appropriate programs in advance, it may be reduced to reflexive reaction to crisis in an atmosphere not unlike that of the Great Depression. Alternatively, and as we will discuss in future pages, the coming transition may necessarily be accomplished outside the traditional framework of a centralized government.

The Genesis Strategy [49]

One of the most useful single steps that industrial governments could take is immediately to establish a massive food-reserve program. In 1961 the world's reserves of grain were sufficient to feed the global population for 105 days. But today the earth's population is larger and food reserves have shrunk. By 1976 grain reserves were sufficient to last only 31 days[50] (Figure 10.7). In the nonindustrial and heavily populated tropical nations of the world, food deficits may double in the coming decade,[51] placing increasing pressure on declining world food supplies. And yet in the U.S. we are evidently blinded by abundance. Each new bumper crop adds to

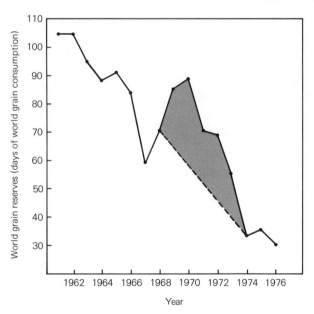

10.7 Decline in a world grain reserves over the past decade. The shaded area under the curve approximately represents the effect of the Green Revolution. (from L. Brown, "The World Food Prospect," *Science* 190 [1975]: 1053–59, Table 1)

the apparently endless complacency with which our nation views the world's food future. In 1977, three successive record wheat crops induced the Carter administration to seek a 20% reduction in wheat acreage for the following year[52] in order to protect farmers' prices. Less than a decade earlier similar reductions in acreage were decreed. As a result, world grain stocks were depleted to levels that were inadequate to cope with the worldwide crop failures of the early 1970s.

But as events of the past furnish due warning, they also suggest an appropriate strategy for the future. In the Book of Genesis, Joseph of Egypt took advantage of seven plentiful years to stockpile food for the famine that followed. For years China and the USSR have pursued national defense policies of building food reserves, and the United Nations has urged a similar policy upon the global community. For years, book after book has documented the need for such a food reserve.[53] There is yet time, for energy and resources are still abundant and comparatively cheap. But as we have seen, the material abundance generated by fossil fuels may soon begin an irreversible decline. If the U.S. is to adopt the ancient Genesis Strategy to prepare for the approaching transition to an energy-scarce future, now must be the time. Instead of continuing to pay farmers huge price subsidies to leave land idle that could otherwise produce food and absorb labor,[54] the U.S. could well stockpile its surpluses and make ready to divert labor to land, in preparation for the possible transition that lies ahead. Hopefully, a way can be found to protect present farm prices without forfeiting the future.

It is of little use for the U.S. to build a food reserve for itself alone, as is implicit in the "lifeboat ethic." This lifeboat that fate has privileged us to occupy is built, propelled, and stocked by energy and resources from abroad. Our dependence on foreign energy and resources may oblige us to continue to share our food with others even if our sense of morality does not. The U.S. comprises but 6% of this planet's population; hence a one-year supply of food for the U.S. is equivalent to only a three-week supply for the entire planet. Joseph of Egypt stockpiled food enough for seven years; the U.S. might well prepare for the approaching transition by setting itself a similar goal.

Looking Ahead

No one can know in advance exactly how our civilization will react to the depletion of energy and natural resources. The present chapter is but a preliminary and speculative analysis of one possible scenario. But regardless of the details of the future, continued survival of the human population will require food; and without abundant energy and natural resources to grow food, increased human labor is the only path that is available.

The need to divert human labor from industry to agriculture is staggering in its implications. Taken to its logical conclusion, it implies no less than an end to the Industrial Age and a comprehensive restructuring of the civilization that now inhabits this planet. In the remainder of this book I will explore these changes by continuing to follow the logical consequences of energy and resource depletion wherever they may lead. We will see that the exhaustion of energy and resources may alter our residential patterns and transform our cities beyond recognition. Energy and resource depletion may revise the purpose and structure of education, medicine, science, government, and the military. And energy and resource depletion may ultimately change for the better the human relationship with nature, and with it our view of self and our position in the universe.

NOTES

1. L. M. Thompson, "Weather Variability, Climatic Change and Grain Production," in *Food: Politics, Economics, Nutrition and Research*, edited by P. H. Abelson (Washington, D.C.: American Association for the Advancement of Science, 1975), pp. 43–49.

2. J. D. Gavan and J. A. Dixon, "India: A Perspective on the Food Situation," in *Food: Politics, Economics, Nutrition and Research*, edited by P. H. Abelson (Washington, D.C.: American Association for the Advancement of Science, 1975), pp. 49–57 (see especially Figure 1, relating weather variability to summer wheat production); S. S. King, "Iowa Corn and Soybeans to Suffer from State's Driest Six Months," *New York Times*, 7 December 1976, p. 22; "Winter Wheat Crop Deteriorates," *New York Times*, 11 February 1976, p. 61; H. J. Maidenberg, "Crop Planting Delays Disrupt Fertilizer Industry," *New York Times*, 5 April 1978, p. 47; R. Lindsey, "Rain Ends Costly California Drought; Cold Now Feared," *New York Times*, 6 February 1976, p. 50.

3. A. Makhijani, *Energy and Agriculture in the Third World* (Cambridge, Mass.: Ballinger, 1975), p. 2; S. Wortman, "Food and Agriculture," in *Food and Agriculture* (San Francisco: W. H. Freeman, 1976), pp. 3–11.

4. F. M. Grodie, *Thomas Jefferson: An Intimate History* (New York: W. W. Norton, 1974), p. 156.

5. T. T. Poleman, "World Food: A Perspective," in *Food: Politics, Economics, Nutrition and Research*, edited by P. H. Abelson (Washington, D.C.: Ameri-

can Association for the Advancement of Science, 1975), pp. 8–16 (quoted data from p. 12).

6. Ibid.

7. F. M. Lappe, *Diet for a Small Planet*, rev. ed. (New York: Ballantine, 1975).

8. Poleman, "World Food."

9. W. Robbins, "Cattleman's Valley Offers No Shelter from Losses," *New York Times*, 20 September 1977, p. 39.

10. B. F. Johnston and P. Kilby, *Agriculture and Structural Transformation* (New York: Oxford University Press, 1975).

11. W. J. Chancellor and J. R. Gross, "Balancing Energy and Food Production, 1975–2000," *Science* 192 (1976): 213–18.

12. Johnson and Kilby, *Agriculture and Structural Transformation*.

13. U.S. Department of Agriculture, Economic Research Service, "The World Food Situation and Prospects to 1985," Foreign Agricultural Economic Report #98, p. 10.

14. D. Pimentel, L. E. Hurd, A. C. Bellotti, M. J. Forster, I. N. Oka, O. D. Sholes, and R. J. Whitman, "Food Production and the Energy Crisis," in *Food: Politics, Economics, Nutrition and Research*, edited by P. H. Abelson (Washington, D.C.: American Association for the Advancement of Science, 1975), pp. 121–27.

15. Ibid.

16. Ibid.

17. Ibid.

18. Ibid.

19. Ibid.

20. U.S.D.A. Economic Research Service, "The World Food Situation."

21. J. S. Steinhart and C. S. Steinhart, "Energy Use in the U.S. Food System," *Science* 184 (1974): 307–16.

22. U.S. Census Bureau, *Pocket Data Book: USA 1973* (Washington, D.C.: U.S. Government Printing Office, 1974), p. 209, table 305.

23. Ibid, p. 201.

24. "Inflation Widened Gap Between Nation's Rich and Poor in 1974," *New York Times*, 2 February 1976, p. 41.

25. "Standard of Living Said to Fall in '75," *New York Times*, 27 December 1976, p. D4.

26. E.g., E. Cowan, "Consumer Prices up by 0.8% for January, Double Average Rise," *New York Times*, 28 February 1978, p. 1; C. H. Farnsworth, "Living Costs up 0.6% in February, Spurred by Food Price Rise," *New York Times*, 29 March 1978, p. 1.

27. R. Lindsey, "Cost of Average Home Tops $50,000," *New York Times*, 23 October 1976, p. 35.

28. E.g., G. F. Sprague, "Agricculture in China," in *Food: Politics, Economics, Nutrition and Research*, edited by P. H. Abelson (Washington, D.C.: American Association for the Advancement of Science, 1975), pp. 57–63.

29. F. Butterfield, "China in Major Drive to Mechanize Farms," *New York Times*, 25 December 1976, p. 1; id., "Peking Party Chief Pledges to Improve Standard of Living," *New York Times*, 27 February 1978, p. 1.

30. S. Wortman, "Food and Agriculture," in *Food and Agriculture* (San Francisco: W. H. Freeman, 1976), pp. 3–11.

31. S. H. Wittiwer, "Food Production: Technology and the Resource Base," in *Food: Politics, Economics, Nutrition and Research*, edited by P. H. Abelson (Washington, D.C.: American Association for the Advancement of Science, 1975), pp. 85–90.

32. Wortman, "Food and Agriculture."

33. Steinhart and Steinhart, "Energy Use in the U.S. Food System"; Pimentel et al., "Food Production and the Energy Crisis."

34. B. Rensberger, "Expert Says Only Hope to Feed World Is with Food Production Unlike That in U.S.," *New York Times*, 8 December 1976, p. A19.

35. J. D. Gavan and J. A. Dixon, "India: A Perspective on the Food Situation," in *Food: Politics, Economics, Nutrition and Research*, edited by P. H. Abelson (Washington, D.C.: American Association for the Advancement of Science, 1975), pp. 49–57.

36. E.g., Johnston and Kilby, *Agriculture and Structural Transformation*.

37. J. Collins and F. M. Lappe, "Still Hungry after All These Years," *Mother Jones*, August 1977, pp. 27–33.

38. Ibid.

39. Ibid.; J. Collins and F. M. Lappe, *Food First: Beyond the Myth of Scarcity* (Boston: Houghton Mifflin, 1977).

40. D. Zwerdling, "The Day of the Locust," *Mother Jones*, August 1977, pp. 35–38.

41. J. R. Harlan, "Our Vanishing Genetic Resources," in *Food: Politics, Economics, Nutrition and Research*, edited by P. H. Abelson (Washington, D.C.: American Association for the Advancement of Science, 1975), pp. 157–60.

42. S. Beart, "Genetic Peril in Cattle," *Environment* 19 (1977): 2–3.

43. Zwerdling, "The Day of the Locust."

44. "Family Farm Held Most Productive," *New York Times*, 10 February 1976, p. 29; P. K. Bardham, "Size, Productivity and Returns to Scale: An Analysis of Farm-level Data in Indian Agriculture," *Journal of Political Economy* 81 (1973): 1370–86.

45. Center for Science in the Public Interest, *From the Ground Up: Building a Grass Roots Food Policy* (Washington, D.C.: Center for Science in the Public Interest, 1975), p. 124.

46. Collins and Lappe, "Still Hungry after All These Years."

47. R. A. Jones, "Efforts to Buy Land in Water District Told," *Los Angeles Times*, 17 February 1976, p. 2.

48. S. S. King, "Mondale Says Carter Plans to Ease Farm Land Limit," *New York Times*, 13 January 1978, p. 1; S. S. King, "Bergland Would End 160-acre Farm Limit," *New York Times*, 27 January 1978, p. A10.

49. S. H. Schneider, with L. E. Mesirow, *The Genesis Strategy* (New York: Plenum, 1976).

50. L. Brown, "The World Food Prospect," *Science* 190 (1975): 1053–59.

51. V. K. McElheny, "Doubling of Food Deficit of Tropical Nations Possible," *New York Times*, 3 August 1976, p. 37.

52. C. H. Farnsworth, "Carter to Seek Cut in '78 Wheat Crop; Food Reserve Asked," *New York Times*, 30 August 1977, p. 1.

53. E.g., Schneider, with Mesirow, *The Genesis Strategy*, L. Brown, *In the Human Interest* (New York: W. W. Norton, 1974).

54. S. S. King, "Senate Notes Rise in Farmer Subsidies for Grain and Cotton," *New York Times*, 22 March 1978, p. 1.

Chapter 11
POPULATING
THE EARTH

One of the most dramatic and symptomatic events of the Industrial Age is the explosive growth of the human population. In thermodynamic terms, the virtual sea of humanity that now populates this globe is a form of prehistoric sunlight, stored over the millennia in fossil fuels and resurrected by industrial agriculture into living human protoplasm. As we have seen, the population explosion may be traced ultimately to the use of energy and resources to accelerate the food cycle. The harnessing of energy and resources is also responsible for one of the most impressive monuments to the Industrial Age, the cities that house the human hordes. The Pyramid of Cheops, the Acropolis, the Temples of the Mayans—all of these physical tributes to earlier civilizations are dwarfed by the endless sea of skyscrapers, glassy pinnacles, and concrete canyons that typify the modern industrial city.

But as we have seen, there is reason to believe that the flood tide of energy and resources with which our civilization has been populated and constructed may be approaching an irreversible ebb. If fossil fuels have been responsible for the growth of human numbers, we are led naturally to wonder whether today's swollen human population can be sustained as fossil fuels enter the period of decline. Likewise, if energy and resources underlie the erection of an urban civilization, we might well inquire whether the decline of energy and resources signals the beginning of its end.

In this chapter we will address the interrelated issues of population and the cities with the aid of the ecological concept of carrying capacity. I will suggest that the earth is in principle capable of supporting its present

215

human numbers in an energy-scarce future, but only with a radical redistribution of population. Indeed, we will see that the depletion of energy
and resources may imply the dismantling of urban civilization as we have
known it.

The History
of Human Population

One of the earliest, best known, and most controversial students of the
human population was the English mathematician Thomas Malthus.
Writing nearly two centuries ago, Malthus pioneered modern population
theory with his famous "postulata," in which he wrote:

> I think that I may fairly make two postulata. First, that food is necessary
> to the existence of man. Second, that the passion between the sexes is
> necessary, and will remain nearly in its present state. . . .
>
> Assuming then, my postulata as granted, I say, that the power of
> population is indefinitely greater than the power in the Earth to produce
> subsistence for man.
>
> Population, when unchecked, increases in a geometrical ratio. Sub-
> sistence increases only in an arithmetical ratio. A slight acquaintance
> with numbers will shew the immensity of the first power as compared
> with the second.[1]

Malthus was in effect saying that the human population increases exponentially, while food production increases linearly. Therefore population growth could quickly be expected to outstrip the food-producing capacity of this planet. What Malthus could not have foreseen is the
exponential growth of food production (subsistence) made possible by the
application of energy and resources to agriculture (Chapter 9). Because
industrial agriculture has kept well ahead of population growth, the ominous implications of Malthus's postulata have not yet been realized. Instead, the story of the human population is one of unremitting expansion.
In the million years before the birth of Christ, human numbers are estimated to have increased slowly to less than one-tenth of one billion (Figure 11.1). But we have seen how the advent of agriculture about 10,000
years ago accelerated the food cycle, with the consequence that human
population growth could also begin to accelerate (Figure 11.1). With the
onset of the Industrial Revolution in the 1700s, the human species began
to apply fossil fuels and natural resources to farming. At last food could be
produced faster than the maximum biological rate of human reproduc-

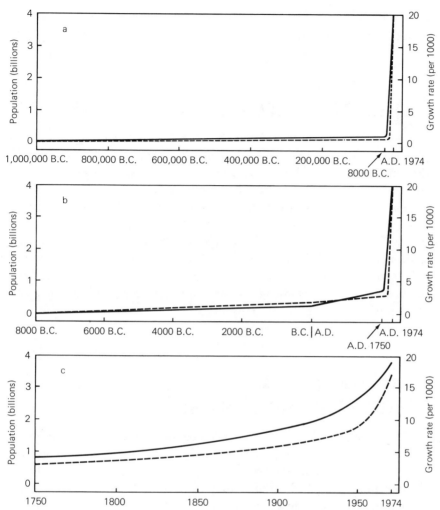

11.1   Growth of the human population over time. The graph in *a* provides an overview, and emphasizes the sharply dichotomous character of human population growth (preindustrial vs. postindustrial). The graph in *b* represents about 1% the time span of *a*, and shows the expansion of human population since the introduction of agriculture. The graph in *c* shows the period since the beginning of the industrial revolution. Solid curves, population size; dashed curves, population growth rate. (from A. J. Coale, "The History of Human Population," *Scientific American* 231 [1974]: 42)

tion. Population growth was released temporarily from its natural limits, raising the gain of the positive-feedback reproductive process well above the magic number 1.0 (Chapter 2). The steeply climbing phase of exponential growth began in earnest, and continues to this day (Figure 11.1). The time from 1750 to the present represents less than 0.02% of

human history and yet 80% of the increase in human numbers has occurred during this period.

Population Growth

What are the forces that underlie the growth of populations? If we sim-
plify the answer by ignoring the potentially substantial effects of migration,[2] population size is governed by two variables: birth rate (natality)
and death rate (mortality). When natality exceeds mortality, population
increases in size. The dramatic increase in human numbers in recent
times is typically attributed to the effects of industrialism on child mortality. As the human species has learned how to grow food faster, death rates
associated with malnutrition have declined, especially among the vulnerable young. At the same time the knowledge generated by industrialism
has furnished the basis of modern medicine and public health, which has
also reduced infant and child mortality.

The Theory of
Demographic Transition

While overall global population has increased with exponential abandon
in the past three centuries, the pattern of growth has varied from one nation to the next, depending largely on the stage of industrialization. From
the beginning of the Industrial Revolution to the 20th century, population
grew fastest in the industrialized nations of the world, as would be expected from the aforementioned effects of industrialism on mortality. But
since 1920 the populations of most industrial nations have stabilized at a
low growth rate, typically under 1% per year (Figure 11.2). In contrast,
the nonindustrial nations have followed a fundamentally different pattern.
At about the same time that population growth leveled off in industrial nations, it began to surge in nonindustrial nations (Figure 11.2). Instead of
industrializing first and growing in population as a result, nonindustrial
nations have been able to import from established industrial nations the
food, materials, and technology that underlie population growth. Thus
the growth of global population that continues even now is causally linked
to industrialism, although the effect is now manifest mainly in the nonindustrial nations.

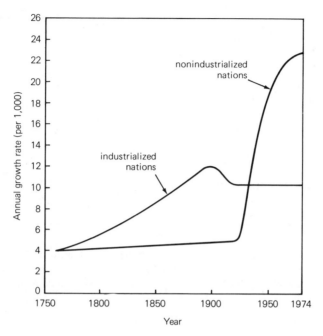

11.2 Population histories of industrialized and nonindustrial nations. The industrialized nations grew first in population, but then leveled off. In contrast, nonindustrial nations grew slowly until after World War I, whereupon their growth rate began to soar. (modified from A. J. Coale, "The History of Human Population," *Scientific American* 231 [1974]: 47)

Demographers—students of human population—have attempted to explain the stabilization of population growth in industrialized nations with the concept of demographic transition. According to this concept, the advances in public health and nutrition that accompany industrialism lower the death rate, especially among children, and population therefore begins to grow. But the rise in per capita income that also attends industrialism then reduces the perceived economic value of bearing several children, and birth rate accordingly declines owing to voluntary contraception. Consequently the initial spurt of exponential growth slows and stabilizes at a new and higher level of equilibrium (Figure 11.3).

The theory of demographic transition has potential implications for the population policies of both industrial and nonindustrial nations. If the theory is correct, the growth of population might be curbed by helping the nonindustrial nations to industrialize, and statements such as "development is the best contraceptive"[3] might have some basis in fact. But the "theory" of demographic transition is in reality an untested hypothesis that lacks supporting data. Evidence is accumulating that fluctuation in

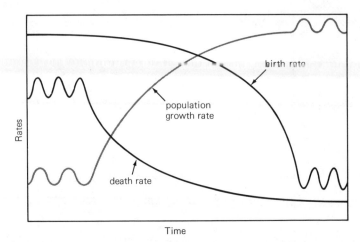

11.3 Schematic illustration of the theory of demographic transition. According to this theory, population growth results from a decline in death rate that is not accompanied immediately by a decline in birth rate. Population thus increases, until curtailed by a later decline in birth rate.

birth rate is the initial and principal cause of population variation, both in "primitive" societies[4] and industrial nations.[5] Even if the theory of demographic transition is valid for industrial nations of the past, the inverted developmental sequence of nonindustrial nations renders the theory questionable as a present population policy. Unchecked population growth within nonindustrial nations acts as a "sponge," absorbing any surplus capital and hence retarding the positive-feedback reinvestment process that has in the past formed the basis of industrialization. The nonindustrial nations are thus caught in a circularity: even if development could stabilize population growth, population growth must stabilize in order to permit development. The theory of demographic transition offers no exit from this dilemma.

The Age Structure
of Populations

Whatever the cause of population stabilization in the industrialized world, the effect is to influence profoundly the age structure of the population. When population growth is rapid, many babies are born per unit time and the "age mix" of the population is weighted toward the young.

The predominance of young is evident in a "frequency histogram" show-ing the percentage of the population in each age group. In growing popu-lations such as the U.S. in 1900 and the nonindustrial nations of today, the histogram is broadest at the base and narrowest at the top, and thus resembles a pyramid (Figure 11.4a). But in stable populations, such as the contemporary U.S., the very young comprise a smaller percent of the populace, as reflected in a rectangular population histogram (Figure 11.4b). Assuming that population growth remains stationary in the U.S. and other industrialized nations for the next two decades, our population in the year 2000 will unavoidably be composed largely of older people (Figure 11.4c). A recent report of the U.S. Census Bureau shows that if present birth trends continue, then by the year 2030 nearly one-fifth of the U.S. population will be over 65 years of age.[6]

The age structure of the U.S. population has major socioeconomic implications for an energy-scarce future. The median income of older families is typically about half that of younger families.[7] Moreover, the aged members of industrial societies are supported largely by fixed retire-ment incomes and hence the aged are among the most vulnerable victims of inflation. As the proportion of old people increases, they may be ex-pected to comprise an ever-greater fraction of people living below the poverty line. Already the aged are increasingly confined by economics to decaying central cities, where the "golden years" are tarnished by inces-sant fear of violence.[8] As inflation is driven relentlessly in the coming de-cades by energy and resource depletion (Chapter 5), the plight of the aged may be elevated to a foremost sociological problem throughout the industrialized world.

The economic implications of an increase in the aged population are likewise far-reaching. As the proportion of people reaching retirement age increases, the already severe economic pressure on pension and retirement funds is bound to escalate.[9] According to *Fortune* magazine, the collective obligation of corporate pension plans in the U.S. may ex-ceed their current assets by as much as $100 billion.[10] Likewise, the U.S. Social Security system has flirted with bankruptcy, and has been kept sol-vent only by a steady increase in Social Security taxes[11] and a decrease in benefits.[12]

Moreover, the increased proportion of old people may imply in-creased political power in the hands of the old,[13] as previewed by the recent increase in the mandatory retirement age in the U.S. from 65 to 70.[14] Such an increase means that a larger fraction of the limited jobs that can be provided by a declining industrial economy will be held by older people, effectively barring the entrance of young people into an already

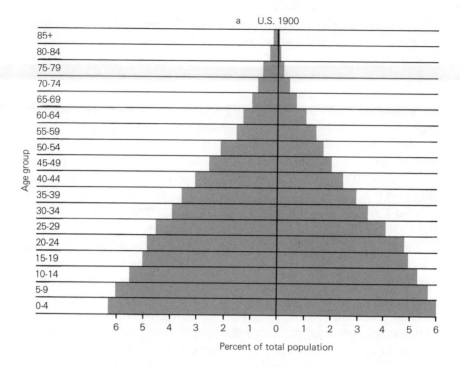

a   U.S. 1900

Age group

85+
80-84
75-79
70-74
65-69
60-64
55-59
50-54
45-49
40-44
35-39
30-34
25-29
20-24
15-19
10-14
5-9
0-4

6 5 4 3 2 1 0 1 2 3 4 5 6

Percent of total population

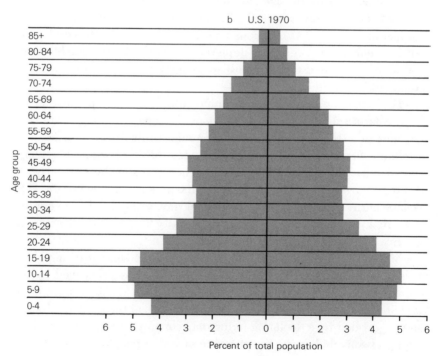

b   U.S. 1970

Age group

85+
80-84
75-79
70-74
65-69
60-64
55-59
50-54
45-49
40-44
35-39
30-34
25-29
20-24
15-19
10-14
5-9
0-4

6 5 4 3 2 1 0 1 2 3 4 5 6

Percent of total population

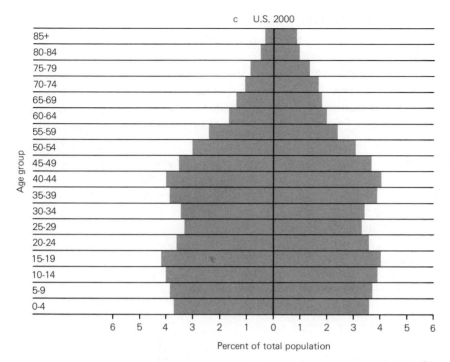

11.4 Frequency histograms for the population of the United States in 1900 (*a*), 1970 (*b*), and 2000 (*c*). Note the increased proportion of older people, implicit in a stationary population growth rate. (from E. J. Kormondy, *Concepts of Ecology*, 2nd ed. [Englewood Cliffs, N.J.: Prentice-Hall, 1976], pp. 114, 116)

restricted job market.[15] As we shall see in a future chapter, this process has already begun, and it may increase the likelihood that the coming transition will take place outside the traditional socioeconomic framework.

The Future of the Global Population

Far from suffering the catastrophic collapse that Malthus might have predicted, the human population has grown without interruption in recent history. For the immediate future, at least, the United Nations World Population Conference of 1974 concluded that "little change is expected to occur in average rates of population growth either in the more developed or less developed regions."[16] In the long term, however, it is self-evident that a finite planet cannot sustain an infinite population, and that population growth rates must eventually approach zero (Figure 11.5). The issue is not whether, but rather how and when, population growth will cease. A prevailing view is that Malthus was not fundamentally in error, but merely premature. According to this view the depletion of energy and resources marks the decline in global food production, with corresponding forced shrinkage in the size of the human population. Proponents of this neo-Malthusian position note that population fluctuation occurs naturally in animal populations, accompanied by significant genetic selection,[17] and that there is no reason to believe our species is exempt from such a pervasive natural law. Indeed, since W.W. II alone, humanity has experienced at least 50 major famines, all caused by crop failure during drought.[18] Inhabitants of industrial nations have been largely insulated from hunger by their affluence, but with the decline of energy and resources such insulation may vanish.

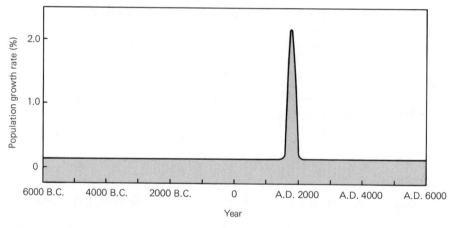

11.5 Rate of human population growth throughout human history and into the future. The rapid growth rate will of necessity represent a transitory phase of human history. Note the similarity of this curve to King Hubbert's hump (Figure 3.8). (after R. Freedman and B Berelson, "The Human Population," *Scientific American* 231 [1974]: 36–37)

The Concept
of Carrying Capacity

There is an alternative to the Malthusian solution, an alternative that may be termed the "steady-state" scenario. By this view, worldwide population will soon reach a sustainable plateau and remain stationary into the indefinite future. Early evidence of such a possible scenario is available in the recent decline in fertility throughout much of the nonindustrialized world.[19] In China, for example, natality has declined an estimated 25% since 1950, with most of the decline occurring since 1965. This population trend may result from a rise in the age at which people marry.[20] If it is sustained, the trend implies that the rate of world population growth has already peaked and is now stabilizing. In this case the global population would rise to 5.5–5.8 billion by the year 2000, and about 10 billion by the end of the next century.

Whether even this more conservative projection is realistic of course depends on the ability of our planet to sustain such a large human population. Ecologists measure this ability through the "carrying capacity," briefly defined as the maximum population density that can be supported by a given environment.[21] In the case of the human species, population density is necessarily an inexact measure of carrying capacity. Because *Homo sapiens* is an intelligent, social animal, such variables as political organization and sociological structure also play a significant role in determining carrying capacity. To illustrate, China and India are comparable in population density, but while China feeds its own populace, India suffers from chronic hunger.[22]

While mean population density is not a precise measure of human carrying capacity, it has the virtue of simplicity. Moreover, it is a fact that each living human being has a more or less fixed daily nutritional requirement: the efficiency of the photosynthetic process that furnishes our food is immutable; and the land area of our planet is effectively stationary. In the most basic ecological sense, human carrying capacity can therefore ultimately be expressed in terms of population density.

Global Carrying Capacity

How large a human population density can our planet tolerate? Estimation of the human carrying capacity of the earth has seldom if ever been attempted, and for good reason. The variables are formidably complex, subjecting even the most careful and informed calculation to wide uncertainty. Fortunately, there is no need to rely on uncertain theoretical calculations, for history provides sufficient empirical examples from which to draw. In populations of Alaskan Eskimos, for example, a sustainable daily harvest of 2,400 calories per person requires an estimated 140 square kilometers of land per person.[23] Such a low population density is mandated by the harsh northerly environment. Overall photosynthetic productivity is low and agriculture is impossible, and hence the Eskimos must obtain nourishment largely from the top of natural food chains. In contrast, the American Plains Indians of more southerly latitudes required only 25 square kilometers of land per person to obtain a daily intake of 2,400 calories.[24]

The practice of agriculture vastly increases the effective carrying capacity of a given environment by acclerating the food cycle. Thus, under the subsistence agriculture practiced by preindustrial societies, a daily nutritional harvest of 2,400 calories required an estimated one square kilometer per person.[25] The application of energy to agriculture is estimated to raise productivity some 100-fold.[26] But as we have seen, "productivity" is not synonymous with "efficiency." Relying on human labor rather than energy, the contemporary Chinese have evolved the most efficient large-scale agricultural system in recorded history. Each Chinese obtains sufficient and sustained nourishment from a plot of land no larger than a backyard garden. A daily intake of 2,400 calories is furnished by only 230 square meters of arable land per person,[27] corresponding to a population density of about 18 people per acre. It is probably safe to assume that this figure represents a practical maximum human carrying capacity, one that is unlikely to be exceeded by any society in an energy-scarce future.

Global Population Densities

To estimate whether the present human population can be sustained when fossil fuels and resources are no longer available, we can compare existing population densities with the practical maximum attained in contemporary China. The total land area of our planet is about 136 million square kilometers, about 10% of which is considered arable (excluding grazing lands, also equal to about 10% of the total). In 1973 the estimated midyear global population was nearly 4 billion people. Hence, the mean human population density of the entire planet was 280 people per square kilometer of arable land (Table 11.1). Such a population density is equivalent to about 3,500 square meters of arable land per person (1.1 persons per acre of arable land; Table 11.1).

Table 11.1. POPULATION DENSITY PER SQUARE KILOMETER OF
ARABLE LAND, SQUARE KILOMETERS OF ARABLE
LAND PER PERSON, AND POPULATION DENSITY PER ACRE OF
ARABLE LAND FOR VARIOUS REGIONS OF THE EARTH [a]

Macroregion or nation	Population per square kilometer of arable land	Approximate square meters of arable land per person	Approximate population density, people per acre
World total	280	3,571	1.1
Africa	120	8,333	0.5
America	130	7,692	0.5
North America	110	9,091	0.4
Latin America	150	6,666	0.6
Tropical South America	120	8,333	0.5
Middle America (mainland)	300	3,333	1.2
Temperate South America	100	10,000	0.4
Caribbean	1,160	882	4.7
Asia	800	1,250	3.2
East Asia	830	1,205	3.4
Mainland	720	1,389	2.9
Japan	2,900	345	11.7
Other	2,540	394	10.3
South Asia	770	1,299	3.1
Middle	1,220	820	4.9
Southeast	680	1,471	2.8
Southwest	190	5,263	0.8
Europe	960	1,042	3.9
Western	1,620	617	6.6
Eastern	1,080	926	4.4
Southern	1,000	1,000	4.0
Northern	500	2,000	2.0
Oceania	20	50,000	0.1
USSR	110	9,091	0.4

[a] Data are based on the simplifying assumption that 10% of the land area in each region is arable, and use 1973 population figures. The available land per person in all cases exceeds the 230 square meters required under Chinese agriculture. Calculated from data in P. Reining and I. Tinker, eds., *Population: Dynamics, Ethics and Policy* (Washington, D.C.: American Association for the Advancement of Science, 1975), p. 80; all original data are described as "estimates of the order of magnitude and are subject to a substantial margin of error."

Similar calculations for each of the earth's major land masses (Table 11.1) reveal that the amount of arable land per person ranges from 1.5 times the Chinese minimum of 230 square meters per person (Japan) to 217 times this value (Oceania). Of the major subregions of the earth, only Japan exhibits population densities comparable to China. Most of Europe, Middle South Asia, and the Caribbean follow close behind, however, each having population densities such that the amount of arable land per person ranges from 2.68 to 3.75 times the Chinese minimum (Table 11.1). It is no accident that the most densely populated regions of the world (Table 11.1) are either highly industrialized and therefore rely heavily on fossil fuels for food (Japan and Europe) or are chronically hungry (parts of Asia and the Caribbean).

The above line of reasoning is clearly unrealistically simplified. It does not take into account regional variations in the amount of arable land, water, photosynthetic productivity, political organization, or cultural history, all of which have a significant impact on effective human carrying capacity. And yet this simplified approach nonetheless suggests that as a first approximation, the human population has not yet exceeded the capacity of the earth to support it. Assuming that agricultural efficiencies comparable to the contemporary Chinese can be implemented on a global basis—a bold assumption at best—then the present human population of the planet can probably support itself even without the help of fossil fuels. Indeed, if arable land were the only consideration, then the earth has sufficient arable land to support about 59 billion people at population densities comparable to China (one person per 230 square meters).

Arable land, however, is not the only consideration. Long before arable land becomes limiting, the limited availability of other primary natural resources—such as water and wood—would presumably constrain the growth of the human population. Indeed, even with a human population of only double today's it seems likely that vast tracts of our planet will contain more people than can be supported in the face of declining energy and resources. There is time yet for the exercise of choice; but unless humanity soon drastically curbs the growth of its population it may expand beyond the earth's effective carrying capacity, leaving the Malthusian scenario as the only solution to the population "problem." By action or inaction, a decision will be reached within the lifetime of most people now alive.

The Urbanization of the Human Population

The simplistic calculus of mean population density suggests that as a first approximation a nonindustrial, labor-intensive agriculture can support the existing human population. But such calculations assume a population that is spread evenly over the earth's arable lands, a circumstance that is far from the reality of modern industrial nations. In the U.S., for example, more than 80% of the populace is now classified as urban (Figure 11.6), and more than half our people live on less than 1% of the land. The U.S. is not alone in this regard; most industrial nations exhibit comparable population distributions, with the majority of the citizenry packed into urban centers that comprise but a fraction of the total land area, and in the nonindustrial nations the rush to the cities is in full swing.[28]

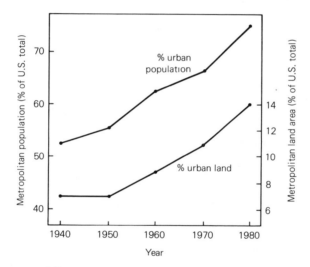

11.6   Urbanization of the U.S. population. Note that the U.S. has become predominantly an urban society (i.e., more than 50% of its population living in cities) only since World War II. (original data from V. P. Barabba, "The National Setting: Regional Shifts, Metropolitan Decline, and Urban Decay," in *Post-Industrial America: Metropolitan Decline and Inter-Regional Job Shifts,* edited by J. W. Hughes [New Brunswick, N.J.: Rutgers University Press, 1975], pp. 39–76, exhibit 1)

The Formation of Cities

We have seen how the construction of the earliest cities was made possible by the adoption of agriculture. The resultant acceleration of the food

cycle generated surplus food, freeing a small percentage of our ancestors
for the task of city building. The social purposes served by the early cities
is not known with certainty, since only their mute ruins remain. Many of
these earliest cities were walled; the term "burg"—suffix to so many European cities—means fortress. Hence the earliest cities may have existed
at least in part to insulate their inhabitants from external forces. The archaeological record suggests that the first human cities may also have
served as spiritual meccas and administrative centers for agriculture, public works, and trade.[29]

The Industrial Revolution

The application of energy and resources to agriculture so accelerated
the food cycle that an increasing proportion of the populace was freed
from agricultural labor. Thus the relative shift in population from countryside to city, or urbanization, did not begin in earnest until the Industrial Revolution attained momentum. Before 1850 no society was predominantly urban. In England, heartland of the Industrial Age, the majority
of people lived in a rural setting until the turn of the present century.[30]
The U.S. has become predominantly urban only since World War II[31]
(Figure 11.6).

Modern Cities

Large-scale urbanization is a relatively recent phenomenon because
the construction and maintenance of cities requires enormous flows of
energy and resources. Indeed, the history of urbanization correlates almost perfectly with industrialization.[32] In the industrialized nations, the
process may be understood as the linkage of reciprocating positive-feedback cycles, driven ultimately by fossil fuels. The liberation of human
labor from agriculture and attendant migration of populations to urban
centers furnished a vast pool of cheap and willing labor for manufacturing.
At the same time, this pool of humanity furnished a massive market for
the goods and services produced by an industrial economy. In the early
stages of industrialization the process of building cities alone served as a
major source of employment and economic growth.[33]

In the industrial nations of the world, cities formed largely because
manufacturing could be prosecuted most effectively in a centralized environment, with specialized and diverse materials, labor, and technology
close at hand. We have seen, however, that the industrial and nonindustrial nations have followed inverse developmental histories. Nowhere are

these differences reflected more clearly than in the structure of cities. In the industrial nations, cities were established decades, sometimes centuries ago, as centers of manufacturing. As a result these cities are shaped like a doughnut, with an old and decaying central city surrounded by concentric rings of newer urban and suburban growth.[34] In many parts of the industrial world, cities have now grown far beyond their original boundaries and merged with growing adjacent urban areas to create a new phenomenon, the metropolis.[35] The cities of nonindustrial nations, in contrast, have formed not as an integral part of industrialization, but rather in response to it, largely as ports of entry and export for trade with the industrial nations.[36] As a result of their sharply distinct history, cities of the nonindustrial world are typically composed of a sparkling new central city replete with glass-and-steel skyscrapers. These central cities are surrounded by concentric rings of ramshackle tin-and-cardboard hovels occupied by the poor, who are drawn to the city's rim in the usually unfulfilled hope of finding employment and a better life.[37]

Urban Population Densities

Regardless of location and history, the single most distinguishing feature of a city is high population density. Indeed, from 1930 to 1950 the Bureau of the Census defined "urban" as a place where 10,000 or more people lived at a density in excess of 1,000 people per square mile.[38] In the 1960s the mean population density of seven major U.S. cities ranged from 7.6 persons per acre (Pittsburgh) to 16.1 persons (Detroit).[39] In terms of carrying capacity, these figures are of limited use, because the majority of urban land is covered over with layer upon layer of concrete, asphalt, and steel, and is inaccessible to agriculture. Open space in cities, on which food could in principle be grown, ranges from 6% (Detroit) to 28% (New York) of the total land area.[40] If we assume that 20% of residential areas also corresponds to "open space," then population densities per acre of arable land range from 29 (New York) to 108 (Detroit) (Table 11.2). As we have seen, the maximum realizable carrying capacity under a highly efficient nonindustrial agriculture is 230 square meters per person, or about 18 persons per acre of arable land. Therefore, even if we exclude the limits imposed by climate, the availability of water, etc., the population density of U.S. cities far exceeds the natural carrying capacity of the land on which they are situated.

Table 11.2. POPULATION PER ACRE OF "OPEN" LAND IN
SELECT U.S. CITIES AS OF 1960[a]

City	Population per acre of open land	Open land per person (square meters)	% of that required for subsistence at Chinese level of agricultural efficiency (230 m²/person)
Detroit	108.3	37.4	16.0%
Pittsburgh	46.4	87.3	37.9%
Philadelphia	83.9	48.3	21.0%
Los Angeles	47.2	85.8	37.3%
Cleveland	44.5	90.1	39.2%
Chicago	53.3	76.0	33.0%
New York	29.2	138.7	60.3%

[a]Open land is defined as institutional open space, parks and recreation, plus 20% of residential area. Calculated from data in C. Abrams, "The Uses of Land in Cities," *Scientific American* 213 (1965): 150–62.

The Metabolism of the City

How can cities survive even though their human populations vastly exceed the carrying capacity of their land areas? The answer, in a word, is energy. A city is a gigantic living organism with a voracious appetite. The metabolism of this organism demands an uninterrupted flow of inputs from external sources. With population densities ranging to more than 100 people per acre of open land, there is not nearly enough agricultural potential within a city to feed its teeming hordes. Virtually all food is therefore imported to cities; an "average" city of a million inhabitants requires a *daily* input of an estimated 2,000 tons of food.[41]

But food is merely first on a list of several essential inputs to the urban environs. The "average" city of one million also imports 9,500 tons of fuels and 625,000 tons of fresh water each day.[42] Most modern cities once existed on nearby energy sources, but today nearly every city in the industrialized world is dependent on distant, and often foreign, oilfields, located in Arctic expanses, Siberia, the deserts of Arabia, and beneath the seven seas. The same is true of all vital resources. Thus the city of the angels, Los Angeles, now depends for survival on water pumped from beneath the Owens Valley, 500 miles to the north. The water is delivered through deserts and across mountain ranges, using a system of aqueducts, pump stations, and storage reservoirs that dwarfs the ancient Roman waterworks. The pattern repeats throughout history and throughout the cities of contemporary industrial civilization: as local sources of fuel, water, and food are exhausted by growing urban populace, cities must reach out farther and farther to fulfill their vital needs.

Energy plays the central role in every aspect of the cities' metabolism. We have examined already how energy magnifies the productivity of agriculture, enabling a few rural inhabitants to grow food enough to feed the urban masses. We have also seen how energy enables this food to be processed and transported to distant urban centers. Our network of highways made of fossil fuels, carries the lifeblood of the city, its trucks propelled by the liquid fuels that are in shortest supply. Water is pumped through the Los Angeles aqueduct using electricity generated largely from fossil fuels. In short, it is energy that provides the irreplaceable link between the cities and their increasingly remote life-support systems.

The Consequences of Carrying Capacity

What will become of cities if and when energy can no longer provide the essential link to remote life-support systems? If vital resources cannot economically be brought to the people who inhabit cities, then survival is likely to compel people to go to the resources. That is, the populace of industrialized nations may be forced by circumstances to redistribute in accord with the natural carrying capacity of the earth. A simplistic consideration of global population density suggests that human numbers have not yet grown beyond the carrying capacity of our planet; but the local, high-density concentrations of human beings that are epitomized by the modern industrial city are inconsistent with an energy-scarce future.

We have noted that more than 80% of the U.S. population is now classified as urban, a proportion that is typical of the industrialized nations. In contrast, only 20% of the population of the typical nonindustrial nation lives in an urban setting.[43] The remainder labor in the earth to provide food. Assuming that energy and resources are truly in decline, then the near future may see the reversal of a three-century trend toward urbanization in the U.S. and the rest of the industrialized world. Our nation may necessarily dismantle many of its cities and deurbanize, toward the ultimate balance of contemporary nonindustrial nations. In other words, the decline of energy and resources may imply the irreversible decline of the city, as is elaborated in the next chapter.

NOTES

1. T. R. Malthus, *An Essay on the Principle of Population* (London: Macmillan, 1798/reprinted 1906), quoted in T. T. Poleman, "World Food: A Perspective," in *Food: Politics, Economics, Nutrition and Research*, edited by P. H. Abelson (Washington, D.C.: American Association for the Advancement of Science, 1975), pp. 8–16.

2. K. Davis, "The Migrations of Human Populations," *Scientific American* 231 (1974): 92–107; C. B. Keely, "Immigration Composition and Population Policy," *Science* 185 (1974): 587–93.

3. Statement made in debate at the World Population Conference in Bucharest, 1974; quoted in M. S. Teitelbaum, "Relevance of Demographic Transition Theory for Developing Countries," *Science* 188 (1975): 420–25.

4. D. E. Drummond, "The Limitation of Human Population: A Natural History," *Science* 187 (1975): 713–21; G. B. Kolata, "!Kung Hunter-gatherers: Feminism, Diet and Birth Control," *Science* 185 (1974): 932–34, N. McArthur, "The Demography of Primitive Populations," *Science* 167 (1970): 1097–1101.

5. H. Frederiksen, "Feedbacks in Economic and Demographic Transition," *Science* 166 (1969): 837–47; Teitelbaum, "Relevance of Demographic Transition Theory for Developing Countries"; W. Petersen, *Population*, 3rd ed. (New York: Macmillan, 1975).

6. "17% of Nation Is Expected to Be Over 65 in Year 2030," *New York Times*, 1 June 1976, p. 1.

7. Ibid.

8. P. Delaney, "Way of Life of Old People Curbed by Fear of Crime," *New York Times*, 12 April 1976, p. 1.

9. C. Kaiser, "Pension Fund Money Running Out for Fire Dept. in New York City," *New York Times*, 25 January 1978, p. 1.

10. "Pensions: A $100-Billion Misunderstanding" (editorial), *New York Times*, 4 January 1978, p. 30; R. Lindsey, "Pension Plans Cancelled by 5,500 Small Companies," *New York Times*, 8 March 1976, p. 1.

11. E. Cowan, "Congress Approves Tax Increases for Financing of Social Security," *New York Times*, 16 December 1977, p. 1.

12. D. Rankin, "Social Security Tax Up, But Future Holds Pension Cut," *New York Times*, 28 January 1978, p. 27.

13. S. V. Roberts, "Gray Power' a New Political Force," *New York Times*, 28 November 1977, p. 1.

14. P. Shabecoff, "Carter Signs Retirement Age Bill; Potential Impact of Law Unknown," *New York Times*, 7 April 1978, p. 1.

15. E.g., J. Flint, "Colleges Fear Later Retirements Will Bar Hiring of Young Scholars," *New York Times*, 24 September 1977, p. 1.

16. U.S. State Department, *U.N. World Population Conference, Bucharest, Aug. 19–30, 1974* (Washington, D.C.: U.S. Government Printing Office, 1974), p. 16.

17. J. H. Myers and C. J. Krebs, "Population Cycles in Rodents," *Scientific American* 230 (1974): 38–46.

18. J. Mayer, "Management of Famine Relief," in *Food: Politics, Economics, Nutrition and Research*, edited by P. H. Abelson (Washington, D.C.: American Association for the Advancement of Science, 1975), pp. 79–83.

19. H. M. Schmeck, "Population Experts Say World Birth Rate Is Declining," *New York Times*, 15 February 1978, p. A15.

20. Ibid.

21. E. J. Kormondy, *Concepts of Ecology*, 2nd ed. (Englewood Cliffs, N.J.: Prentice-Hall, 1976).

22. For a discussion of the limitations of population density as a sociological variable, see W. Petersen, *Population*, 3rd ed. (New York: Macmillan, 1975), pp. 476–85; and A. T. Day and L. H. Day, "Cross-national Comparison of Population Density," *Science* 181 (1973): 1016–23.

23. G. Leach, *Energy and Food Production* (Washington, D.C.: International Institute for Environment and Development, 1975).

24. Ibid.

25. Ibid.

26. Ibid.

27. Ibid.

28. E.g., J. M. Markham, "Flight to Cities Empties Spain's Countryside," *New York Times*, 29 March 1978, p. 2; F. C. Turner, "The Rush to the Cities in Latin America," *Science* 192 (1976): 955–62.

29. E.g., Petersen, *Population*; B. Ward, *The Home of Man* (New York: W. W. Norton, 1976); G. Sjoberg, "The Origin and Evolution of Cities," *Scientific American* 213 (1965): 54–62.

30. K. Davis, "The Urbanization of the Human Population," *Scientific American* 213 (1965): 41–54.

31. V. P. Barabba, "The National Setting: Regional Shifts, Metropolitan Decline, and Urban Decay," in *Post-Industrial America: Metropolitan Decline and Inter-Regional Job Shifts*, edited by G. Sternlieb and J. W. Hughes (New Brunswick, N.J.: Rutgers University Press, 1975), 39–76.

32. R. A. Mohl, "The Industrial City," *Environment* 18 (1976): 28–38.

33. E. C. Kirkland, *Industry Comes of Age: Business Labor and Public Policy, 1860–1897* (New York: Holt, Rinehart and Winston, 1961).

34. N. Glazer, "The Renewal of Cities," *Scientific American* 213 (1965): 194–208.

35. H. Blumenfeld, "The Modern Metropolis," *Scientific American* 213 (1965): 64–89.

36. Ward, *The Home of Man*.

37. N. K. Bose, "Calcutta: A Premature Metropolis," *Scientific American* 213 (1965): 90–105; L. Rodwin, "Ciudad Guayana: A New City," *Scientific American* 213 (1965): 122–33; J. Kandell, "Behind a Facade of Luxury, the Cities of Latin America House Their Poor in Squalor," *New York Times*, 4 November 1976, p. 3.
38. Petersen, *Population*, p. 470.
39. C. Abrams, "The Uses of Land in Cities," *Scientific American* 213 (1965): 150–62.
40. Ibid.
41. A. Wolman, "The Metabolism of Cities," *Scientific American* 213 (1965): 178–90.
42. Ibid.
43. A. Makhijani, *Energy and Agriculture in the Third World* (Cambridge, Mass.: Ballinger, 1975).

Chapter 12
THE DECLINE
OF THE CITY

Twentieth-century civilization is based on cities. The Industrial Age has witnessed a steady flow of humanity from countryside to city, and as recently as 1965 an eminent urban sociologist predicted that by 1990 more than half the world's people would probably live in cities of 100,000 or more.[1] Given the momentum of events, he may be right; but as we have seen, the modern industrial city is in fact a vulnerable institution, desperately dependent on waning supplies of energy and natural resources. Cities are built, maintained, fed, watered, and cleansed using energy and natural resources. As was discussed in the preceding chapter, cheap and abundant energy and resources have enabled city dwellers to exceed natural carrying capacities and live at population densities that nature could not otherwise tolerate. And now, as the era of cheap energy draws to a close, the media proclaim the growing distress of the industrial city. What is the nature of this distress? How is it connected to the decline of energy? And what does it portend for the future of the city? I will attempt to answer these questions in this chapter; in doing so I will develop the position that the large-scale industrial city is swiftly approaching obsolescence.

The Plight of the City

The headlines pay almost daily tribute to the "plight" of the American city, dramatized by the near bankruptcy of New York. What is meant by the plight of the cities? The question is most easily addressed using examples of cities in the northeast United States such as New York, Boston, and Detroit. But as I shall document, the pattern that emerges is a general one, visible to some degree in most of the world's industrial cities.

Urban Exodus

In the mid-1960s, at the height of material prosperity, most scholars from diverse disciplines expected the flow of humanity into cities to continue into the next century. Cities had grown steadily throughout the preceding century. In the three decades from 1940 to 1970 urban population increased at a rate in excess of 20% per decade.[2] But in 1970, at the end of these several decades of growth, the aging process began to catch up with the industrial city and at the same time the real price of energy began to increase for the first time in modern U.S. history. I believe it is more than coincidence that the early 1970s also marked the stunning and largely unforeseen reversal of city growth. For the first time in recent history people began leaving cities faster than they arrived. From 1970 to 1974 alone the net population of America's central cities declined by 10%,[3] and the trend shows no sign of abating.[4]

Characteristics of Outmigrants

What kind of people are forsaking city life? In St. Louis, the number of young white adults decreased by 50% from 1960 to 1970, and at the same time the population of elderly whites increased from 14.5% to 19.2%.[5] Given recent fertility trends, the white population of St. Louis will now continue to shrink whether or not the outmigration of whites continues. St. Louis may be an extreme case, but a similar population trend is visible in virtually all large American cities. National census figures show that between 1970 and 1974 the U.S. white urban population declined by 5%, while the black urban population increased by 6%.[6]

The people leaving cities are thus predominantly young and white. They are also comparatively affluent. The mean annual income of families who left cities in the period 1970–1974 was $14,200, contrasted with $12,000 for families who entered, resulting in a net loss of income of $29.6 billion.[7] As a result of these migration patterns, U.S. cities are increasingly becoming the habitat of the old, the black, and the poor. Those who remain in cities frequently do so less from choice than economic circumstance; a recent Gallup poll indicated that 36% of city dwellers would move out if they could.[8]

Job Losses

Cities are losing not only their inhabitants but also their jobs. From 1970 to 1973, the 13 major metropolitan areas in the northeast, north central, and western U.S. grew more slowly than the U.S. average in terms of nonfarm payroll.[9] At least three urban areas, all located in the northeast U.S., have suffered a net decline in the number of jobs available. New York City alone lost well over half a million jobs from 1970 to the present; the annual loss is near 80,000.[10]

What types of jobs are disappearing from the industrial city? From 1953 to 1973, New York lost about 38% of its manufacturing employment.[11] In fact the entire mid-Atlantic and north central U.S. is losing manufacturing jobs, while the South is gaining (Figure 12.1). Over the same time period, jobs in the service industries, including government employment, increased substantially (Figure 12.2). Data such as these are frequently cited to support the emergence of postindustrial service economy. In fact, the increase in service jobs has not nearly compensated for the loss of manufacturing, with a resultant net loss of employment. Moreover, as we have discussed (Chapter 5), and as we shall soon document, the service industries are among the most vulnerable in periods of economic decline.

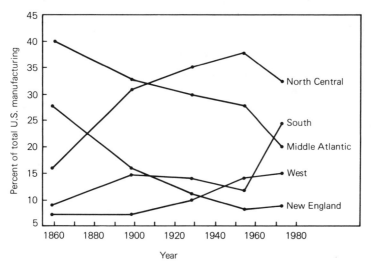

12.1 Regional shift in manufacturing within the United States. The northeastern U.S. has steadily lost manufacturing employment in recent years, while the western and especially southern U.S. have gained. (original data from G. Sternlieb and J. W. Hughes, "Is the New York Region the Prototype?" in *Post-Industrial America: Metropolitan Decline and Inter-Regional Job Shifts*, edited by G. Sternlieb and J. W. Hughes [New Brunswick, N.J.: Rutgers University Press, 1975], pp. 101–37, exhibit 9)

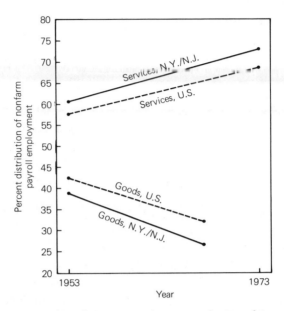

12.2 Manufacturing (goods) and services employment in the United States over the past few decades. Manufacturing has shown a relative decline in the U.S., and even more so in the New York / New Jersey area, while services have shown a relative increase. (original data from G. Sternlieb and J. W. Hughes, "Is the New York Region the Prototype?" in *Post-Industrial America: Metropolitan Decline and Job Shifts*, edited by G. Sternlieb and J. W. Hughes [New Brunswick, N.J.: Rutgers University Press, 1975], pp. 101–37, exhibit 7)

Why is manufacturing leaving the Northeast? There are many reasons, but an important one is the sheer age and technical obsolescence of existing manufacturing facilities. The cities of the northeast U.S. were built in the early stages of U.S. industrialism, and much of the capital equipment used to manufacture goods, as well as the buildings that house this equipment, are in a state of advanced physical deterioration. If energy and natural resources were abundant and therefore cheap, it would in many cases be profitable to refurbish the buildings and replace and modernize capital stock. But energy and resources are no longer abundant nor cheap. Operating on the principle of maximizing profits, manufacturing concerns often simply find it more advantageous to relocate rather than to rebuild.

The High Cost of City Life

Not only is it unprofitable to rebuild and replace deteriorated capital stock, but owing to the high cost of energy, it is increasingly unprofitable to pursue manufacturing in major U.S. cities. Since the early 1970s when

energy prices began their unbroken ascent, the cost of energy has assumed increasing importance to a region's competitive strength in attract-
ing industry. The northeastern U.S., especially dependent on imported
fuel, experienced a 30% increase in energy prices in the early 1970s. In
contrast, cities of the southern U.S., which are located closer to domestic
oil and natural gas supplies, experienced a smaller price increase, ranging
from 7.4% (Atlanta) to 15.4% (Houston).[12] The New York region now has
the most expensive natural gas and electricity in the United States[13] (Figure 12.3).

As we have seen, energy underlies virtually everything in the urban
lifestyle; thus, when the cost of energy goes up, the cost of living follows
faithfully (Figure 12.3). A high cost of living in turn exerts steady upward
pressure on wages, cutting further into manufacturing profits and helping
to drive industry away from the city. It is probably more than chance that
the five urban areas with the highest cost of living in 1973 all lost population between 1970 and 1973, while four of the five regions with the lowest
cost of living gained population during the same time interval.[14] New
York and Boston are the acknowledged intellectual, cultural, and financial
capitals of the U.S., but their residents pay premium prices for these
amenities.

Fiscal Strain
and Positive Feedback

As Muller has noted, the exodus of young white affluents from cities
does not reduce crime levels; it does not reduce welfare rolls; it does not
lessen the number of buildings requiring fire protection; and outmigration from cities does not reduce the miles of streets and highways, and
sewer, power, and water lines requiring maintenance.[15] Because these
costs are relatively fixed, the price of city services does not decline when
people leave the city. Instead, fewer residents must pay for the same services and the per capita tax burden is thus driven steadily upward by the
urban exodus. The taxes paid by New York residents, to cite but one example, are now the highest in the U.S., comprising 10%–20% of the annual family budget.[16] Larger cities again suffer a competitive disadvantage because it costs more money to provide the same service in a large
than in a small city.[17]

The exodus of people from a city marks the beginning of a positive-
feedback cycle of urban decline (Figure 12.4). As people leave, taxes
increase; but as taxes increase, the cost of living increases, driving still
more people from the city. The city has two major alternatives to raising

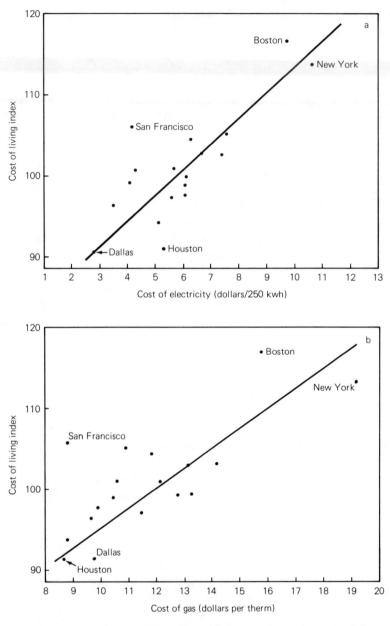

12.3 Relation between the cost of living index and the cost of natural gas (*a*) and electricity (*b*) in major U.S. cities. The strong correlation suggests that the cost of energy is a major contributor to the cost of living. (original data from G. Sternlieb and J. W. Hughes, "Is the New York Region the Prototype?" *Post-Industrial America. Metropolitan Decline and Job Shifts*, edited by G. Sternlieb and J. W. Hughes [New Brunswick, N.J.: Rutgers University Press, 1975], pp. 101–37, exhibits 15 and 16)

taxes: decreasing city services or borrowing money. Decreasing services has been widely employed by city governments; throughout the industrialized world, many cities are simply making do with less.[18] But the strategy is counterproductive because decreasing services lowers the "quality of life," accelerating the exodus of urban dwellers and thereby reinforcing the positive feedback cycle of decline. It is easy to understand why cities have opted for borrowing money.

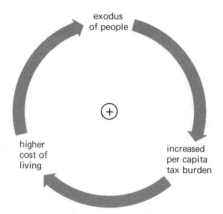

12.4 Schematic illustration of the positive-feedback cycle of urban decline. The exodus of urban population contributes to a higher cost of living for those who remain, who are thus prompted also to emigrate. Higher energy costs fuel the cycle by driving up the cost of living.

In the short run, borrowing money would appear the least painful of the alternatives open to city governments. But in the long run it is the most devastating, and in fact lies at the heart of the immediate plight of the cities. The interest payments on borrowed money, or "debt service," now comprise an average of 7% of all operating costs in larger U.S. cities, and the figure is on the rise. Once again, larger cities are selectively disadvantaged, for debt service represents repayment of previous capital outlays for city services, and as we have seen, per capita service costs are higher in larger cities.[19] Hence per capita debt service follows the same pattern and is higher in larger cities.[20] New York, to cite an extreme case, spent one-fifth of its 1977 fiscal year budget on debt service.[21]

As a city's indebtedness increases, its financial posture weakens and it eventually loses access to credit markets. As a consequence the city can no longer borrow money. Finally, the city can barely meet its staggering financial obligations while maintaining reduced city services. Caught in the positive-feedback cycle of urban decline, the city is driven to the edge of bankruptcy, dramatized most recently by the example of New York.

Generality of the Plight

As the largest and most influential city in the U.S., New York has received special attention from the news media, apparently inducing some to believe that the condition of New York is unique. The Nobel Prize-winning economist Milton Friedman, for example, has written that "New York is a special case. . . . New York's lavish spending reflects the most welfare-state oriented electorate in the U.S."[22]

Is New York in fact a special case or are its well-publicized difficulties symptomatic of a more general malaise? To answer this question, a group of sociologists at the University of Chicago analyzed the "fiscal strength" of more than 50 major U.S. cities, using a barrage of economic indicators.[23] They found that the percentage of the populace on welfare in New York in 1974 was 12.4%, sixth among the cities examined. Ahead of New York in welfare payments were Boston (16.9%), Baltimore (16.3%), Philadelphia (16.2%), St. Louis (15.8%), and Newark (14.4%).[24] Moreover, only 8% of New York's welfare budget was paid from municipal resources; the remaining 92% came from state and federal funds. New York ranks eighth among major cities in the number of ineligibles on welfare, after cities such as Chicago, Boston, and Washington, D.C.[25] In terms of per capita expenditures on city services, Washington, D.C., far outspent all other cities, followed by Palo Alto, California, and then a group of major cities that includes New York (Figure 12.5). These data do not support the contention that New York is the most lavish of the major U.S. cities in its spending practices.

The Chicago sociologists did find that New York has the highest per capita long-term debt of any city studied (Figure 12.6a), and the second-highest per capita short-term debt (Figure 12.6b). But they also found that New York makes a better "tax effort" than most major cities (Figure 12.7). Moreover, New York enjoys a per capita taxable property value of $13,657, one-third larger than the U.S. average of $10,470.[26] Since New York thus has more valuable real estate to draw on as a tax base, it can afford a higher debt than most American cities.

These fiscal facts demonstrate that New York is not alone in its financial problems; rather, it simply inclines toward one end of a continuous spectrum. Other northeastern cities are in equal or greater long-term trouble. Philadelphia experienced a deficit of $100 million in 1975–1976.[27] Much of Detroit is in a state of advanced fiscal and physical deterioration.[28] Cleveland can barely continue certain municipal operations owing to lack of funds.[29] Even the "booming" cities of the South and Southwest are beginning to encounter economic difficulties.[30]

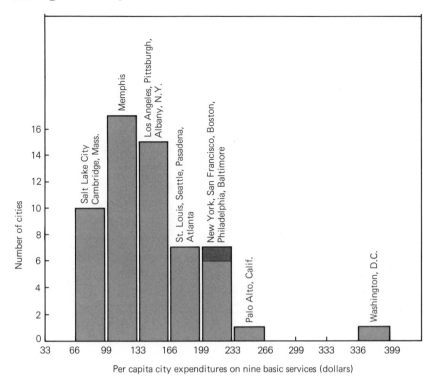

12.5 Per capita city expenditures on nine basic services for 58 major U.S. cities (some of which are indicated by name). The nine services included are police, fire, sewerage, sanitation, highways, parks and recreation, financial, administration, and general control and building. Note that Washington, D.C., far outspent all other cities. (original data from T. N. Clark et al., "How Many New Yorks? The New York Fiscal Crisis in Comparative Perspective," Research Report #72 of the Comparative Study of Community Decision Making [available from T. N. Clark, Dept. of Sociology, University of Chicago, Chicago, IL 60637])

The sickness of the city is not confined to America, but is endemic to all of industrial culture. Some Canadian cities are in decline.[31] Across the Atlantic, the city of Glasgow, built on a now-exhausted coal seam, is in a state of advanced and apparently irreversible deterioration.[32] London is shrinking at an annual rate of 50,000 people, and its inner city is now reduced to two-thirds its 1939 population.[33] In 1973 the population of Paris declined for the first time in its recorded history.[34] On the other side of the industrialized world, ten Japanese cities have declared bankruptcy. Tokyo—the largest city on earth, and capital of the most urbanized nation on earth—is losing population at the rate of 170,000 per year[35] and stands at the brink of default.[36] Even in the partially industrialized USSR, hastily erected cities are aging rapidly, portending an era

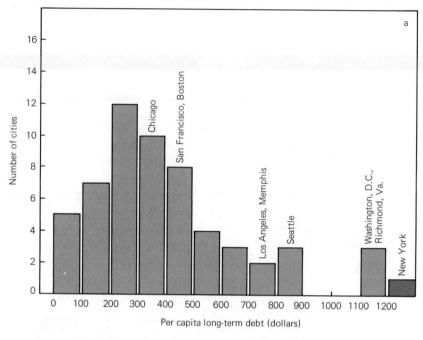

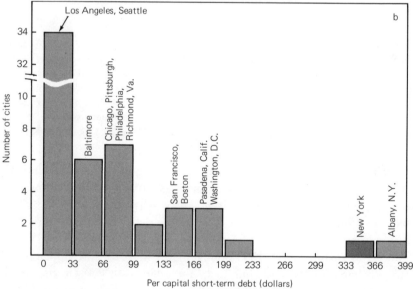

12.6 Per capita long-term debt (*a*) and short-term debt (*b*) of 58 major American cities (some of which are indicated by name). (original data from T. N. Clark et al., "How Many New Yorks? The New York Fiscal Crisis in Comparative Perspective," Research Report #72 of the Comparative Study of Community Decision Making [available from T. N. Clark, Dept. of Sociology, University of Chicago, Chicago, IL 60637])

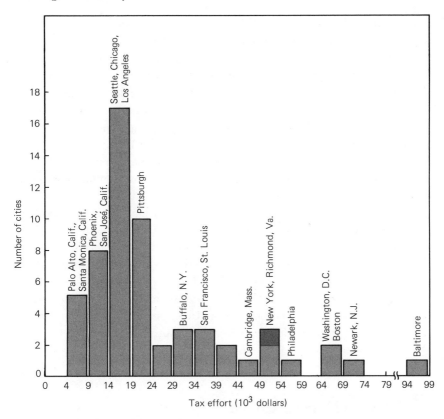

12.7 Tax effort of 58 major American cities (some of which are indicated by name), defined as the tax revenues generated per $1,000 of taxable property value. Note that New York makes a better tax effort than the majority of cities examined. (original data from T. N. Clark et al., "How Many New Yorks? The New York Fiscal Crisis in Comparative Perspective," Research Report #72 of the Comparative Study of Community Decision Making [available from T. N. Clark, Dept. of Sociology, University of Chicago, Chicago, IL 60637])

of physical decline.[37] It seems fair to conclude that the decline of the city is a general phenomenon throughout industrial civilization.

The Suburbanization of America

Where do people go when they leave cities? According to one theory, urban emigrants forsake one city only to erect another. In the U.S., for example, the drain of people from the cities of the Northeast could contribute to the formation of new cities in the South and West. This theory of "reurbanization" thus interprets the present exodus from cities not as decline but as rebirth, part of a natural cycle of urban renewal

in a mature industrial culture.[38] It is important to examine this argument
with care; if it is correct, it implies that the city as an institution may be
alive and well, even though particular cities are in obvious trouble.

There is a simple test of the theory of reurbanization; if it is correct,
then the outmigrants from declining cities ought to end up in new cities.
Census Bureau data in fact indicate that urban outmigrants end up not in
new cities, but largely in the suburbs, continuing a trend that began in
the 1920s with mechanized transportation.[39] At the 1970 census, the population of the American suburbs exceeded that of central cities for the first
time,[40] and by 1974 some 57% of the nation's urban population lived in
suburbs rather than central cities.[41] In the South, which has been attract-
ing northeastern emigrants, metropolitan areas have gained in population, but growth was confined to the suburbs.[42] Only in the western U.S.
has there been any increase in the population of central cities, and even
this increase has been small. Similar trends are evident in Japan, where
virtually every medium-to-large city is losing population to concentric
rings of surrounding suburbs.[43] Now, even the half-century trend toward
suburbanization may be ending in the U.S.; since 1970 rural areas and
small towns have grown faster (6.6%) than metropolitan areas (4.1%) for
the first time in modern U.S. history.[44]

The Human Cost
of Urban Decline:
A Real-Life Scenario

Urban outmigration, the decline in manufacturing, fiscal erosion—these
are among the immediate causes of the city's plight. But these are abstract
concepts, clothed in dry statistics. The heart and soul of a city are its people; the effects of urban decline are revealed most clearly in their daily
suffering. Let us explore the human impact of urban decline, not with
charts and figures, but with a synthesis of firsthand documentary reports
of life in a declining city—in this case, New York.

The Economic Impact

To initiate the cycle of decline, businesses and manufacturers depart
for greener pastures, slowly at first but then in a growing avalanche.[45]
With the jobs go people; the decline in population implies a smaller
market for businesses that remain, which suffer and close in ever-greater

numbers.[46] People are thrown out of work in record numbers,[47] and their life plans radically altered by the turn of events. The city cannot help its inhabitants; its weakened financial posture[48] prevents borrowing more money.[49] State and federal governments make temporary arrangements to fill the city's cash void,[50] but only at the expense of increased external control over the city's internal affairs.[51] Faced with the choice of driving away more businesses with higher taxes or slashing city payrolls and services, city officials choose the latter course,[52] resulting in still greater unemployment. Responding to the cutbacks, local unions that control such vital municipal functions as transportation, garbage collection, building maintenance, and health services walk off their jobs, adding to the atmosphere of crisis.[53] The strikes are disruptive, but do not achieve their economic goal: there is simply no money available.[54] Wages are lost, and accordingly consumers spend less money, impairing the local economy still further as the positive-feedback cycle gains momentum.

The Social Impact

City officials, distressed by the decline, take drastic steps to arrest the slide, including tax advantages to businesses and the relaxation of environmental standards.[55] Both actions are well-meaning attempts to stem the exodus, but they have the opposite effect. The gap between the rich and poor grows larger, pollution increases still more, and that elusive "quality of life" deteriorates further as the tension grows. Crime rises as the poor become both desperate and angered;[56] with no stake in the society, they have nothing to lose by rejecting it. Unemployment among some minority groups rises to 40%–50%, far surpassing depression levels,[57] and a new phenomenon emerges: roving youth gangs in pursuit of subsistence, excitement, and revenge.[58] City councils and police departments react forcefully, and an atmosphere of martial law descends on portions of the central city.[59] Family structure is broken by the strain, especially among the poor, and the number of young urban nomads—children whose homes are the streets—grows into the tens of thousands.[60] Teenagers turn increasingly to prostitution as a means of survival.[61] Funds that furnish unemployment benefits are depleted,[62] fueling the growing fiscal and social chaos.

Fully three-fourths of the city's budget cuts occur in the area of human services.[63] As manufacturing employment declines, there is no longer an economic foundation for maintaining services (Chapter 5); the hope for the post-industrial service economy is thus shattered in the cold reality of urban decline. Children, the old, women, and people with dark skin suffer most from the reductions,[64] as day-care centers are closed,[65] adult education programs ended,[66] schools and universities cut back and are sometimes shut down by lack of operating funds[67] and summer-job programs for youth abruptly terminated.[68] Parks and recreational facilities are curbed,[69] consumer protection agencies abandoned,[70] and institutions of art and music closed[71] as the city's purse strings draw ever tighter.

Human health declines along with the physical condition of the city. Health services are discontinued and hospitals closed.[72] Poorer neighborhoods are particularly hard hit, and are often virtually devoid of doctors and medical personnel[73] as the medical profession follows the economic imperative and moves to the richer suburbs. Mass transit—portrayed only a decade ago as the wave of the future—in practice pays neither its initial capital outlay nor even its day-to-day operating costs.[74] Studies show that even if transit ridership in the nation's cities were doubled, the use of autos would be reduced only marginally.[75] In fact, the transit ridership plummets as the city and its economy declines,[76] requiring a steady reduction in services.[77] The poor, who cannot afford to own automobiles, rely heavily on mass transit. The deterioration of transportation services imprisons them increasingly in their delapidated ghettos, without doctors, without hospitals, and increasingly even without grocery stores as crime forces their closure.

The decline strikes close to home, with frequent interruption of essential serivces such as gas, water, garbage collection, electricity, and heat.[78] Owners of apartment buildings claim that they cannot operate at a profit without increasing rents;[79] and so even with rent controls in effect, rents increase three times faster than incomes.[80]

The Physical Signs

Finally, the surest sign of a civilization in decline appears: block after block of deteriorating buildings are abandoned by their owners, often while people still live within their crumbling walls.[81] These decrepit buildings have on occasion collapsed and crushed their occupants. In the summer of 1976, some 6,800 abandoned buildings were known to New York City officials, who acknowledged that the number was conservative.[82] Eventually whole neighborhoods are abandoned, leaving behind a ghostly urban skeleton, "ravaged sentinels of blight," their silent vigil punctuated by occasional gunshots, the scurrying of rats, and the frequent wail of sirens.

The Cycle Repeats

As the cycle of urban decline gathers force, property values plummet. In New York City, to cite one example, the value of taxable property declined in 1976 for the first time in 33 years.[83] Consequently, tax revenues available to city government drop still further. Unable to borrow more money, the city slashes still more services, and the cycle of urban decline refuels itself in an endless sequence that eventually culminates in bankruptcy.

Urban Bankruptcy

New York City is the intellectual, cultural, and financial capital of the U.S., and no one knows the possible consequences of its bankruptcy. At least one example is available in U.S. history, however—the Massachusetts township of Fall River, which defaulted in 1930. In the early 1920s, Fall River was the cotton-mill capital of the world. During the 1920s, however, the textile industry moved from the northeast U.S. to the South, presaging the subsequent manufacturing shift which we have discussed (Figure 12.1). During the twenties, Fall River failed to reduce expenditures and borrowed vast sums to compensate. Finally the city was unable to pay its bills and slid down the wave of the Great Depression into bankruptcy. In what followed, the State of Massachusetts in essence assumed full financial control over the city, enforcing extreme frugality from a comfortable distance.[84]

Forecasting the Future of the City

Will New York and other American cities repeat the history of Fall River and default in the coming decades? The parallels are striking, and the signs have been evident for some time. In the late 1950s, for example, a nine-volume classic in urban forecasting took note of the decline in manufacturing in the New York area and concluded that "there is no prospect, short of some new force of major dimensions, that this incipient decline will be arrested in the decades ahead."[85] The report suggested that the service industries would take up the economic slack created by the departure of manufacturing, but the recent history of New York and other metropolitan areas would seem to furnish conclusive evidence that a healthy service economy must rest on a sound manufacturing base (Chapter 5).

Much has happened between the late 1950s and the present, including the realization of Hubbert's early predictions of impending energy
scarcity. Modern students of urban affairs are understandably cautious,
but a growing consensus is captured in the following quotations of ur-
banologists:

> We must find a way to admit, politically, that depopulation is not only a
> likely and logical consequence of the aging of urban capital, but is actu-
> ally occurring and demands explicit policy. [86]

> The New York region, once considered to be most stable, has been af-
> fected beyond the capability of the current political system to effect
> repairs in the near future. [87]

> The sharp rise in the cost of energy is likely to encourage further move-
> ment of jobs and households from aging areas . . . to energy rich
> states. . . . [88]

> The older metropolis cannot reverse most of the conditions causing its
> fiscal problems. [89]

> Without the impetus of demand due to population growth, much more
> extensive abandonment of individual houses and whole neighborhoods is
> a likely prospect. [90]

> The recent crises of New York City may well be prototypical of those fac-
> ing other declining metropolises, as well as states and regions. [91]

> Governmental jurisdictions may well have to accept as a fact of life the
> permanence of decay. [92]

> We have yet to feel the full impact of the enormous change . . . of
> energy costs. [93]

> The cities we know, the kind that have provided the physical matrix for
> the flowering of civilizations, are doomed. [94]

Energy, Resources, and National Urban Policy

It is at best simplistic to claim that cities are suffering their present
decline solely because energy and resources have become more expensive. The decline of the city as an institution has many causes, some technical, some sociological, and some no doubt unknown at present. Indeed,
the decline of the city could be claimed with some justification to have its
roots in the early part of this century, with the decentralizing influence of

modern transportation and communication technology. There is no doubt that the decline of the city has many causes; but among them the increase in the price of energy and resources stands out with special relief, heightening a host of old problems and creating new ones. Expensive energy encourages the relocation of manufacturing, helping to precipitate job losses; expensive energy disproportionately increases the cost of urban living, accelerating the exodus of population. Expensive energy causes economic dislocation (Chapter 5) which prevents the rebuilding of neighborhoods, helps keep taxes high, and generates fiscal strain. If energy and resources were cheap and abundant, many of these problems could be relieved. But energy and resources are no longer cheap, and they may not be abundant for much longer. In the most immediate sense, then, the decline of energy and resources may be considered the prime mover—or at least the most proximate precipitator—of urban deterioration.

Cities represent the fullest flower of industrial civilization. They can be architecturally magnificent. They are the repositories of past and present art treasures, the focus of music and drama, the site of the world's finest libraries and universities. Cities are vibrant, living manifestations of the full range of human emotion and accomplishment. Above all, cities are people, their homes, their families and friends, their successes and failures, their past and present lives, and their hopes for the future. The vast majority of people in the U.S. live in an urban setting and are influenced directly or indirectly by the condition of the cities. A national urban policy of "abandoning" the cities is unthinkable. But it is illusion to believe that our cities can be rebuilt, at least on the same scale and organization as in the recent past. If energy and resources are truly in decline, then the age of the city as we have known it is nearly over. In this case an urban policy of rebuilding cities in the old mold is worse than futile—it is tragic. Attempts to rebuild America's decaying cities on the pattern of the past can only nurture false hope, create unfulfilled expectations, divert energy from productive paths, and delay an even more traumatic day of reckoning. In an era of declining energy and resources, the first and essential step toward a useful national urban policy is to accept the decline of the city for the fact that it is. Only then can the climate exist for realistic urban policies.

The city is a dying institution. Like a dying person, it is in need of material aid and spiritual comfort through the trials of a vast transition. The most urgent and immediate urban need is for jobs. Moral issues aside, the widening gap between the rich and poor in cities is likely eventually to precipitate serious social upheaval. A growing population of restless, unemployed, and disadvantaged minorities trapped in ghettos where basic human services are nonexistent poses a clear and present danger to the existing social fabric. Unless meaningful jobs and minimal services are available to those in need, the urban youth gangs of today could become the nuclei of urban guerrilla armies of tomorrow. The limited urban job program initiated recently by the federal government[95] is a hopeful but minute first step. Urban farm projects of the kind co-sponsored by Cornell University and the U.S. Department of Agriculture[96] represent an enormously important social experiment, one that could at once provide meaningful jobs, recycle abandoned urban land, produce wholesome and inexpensive food for hungry city dwellers, and increase the human carrying capacity of the urban environment. Such an innovative experiment in urban farming could be fruitfully expanded and integrated with a national urban job program as the mainstay of an effective short-term urban policy.

In the longer term, however, industrial civilization must come to grips with the need to dismantle the majority of its cities. Of course, the trillions of dollars of past investment that are represented in the physical plant of America's present cities must be used to the fullest extent possible in the coming decades. This economic necessity alone imposes a substantial lead time on the transition away from an urban civilization. But if energy and natural resources are truly in permanent decline, then an urban policy of growth or even steady state is maladaptive. What is required instead is an urban policy of continual retrenchment.

We have seen (Chapter 10) that the depletion of energy and resources implies the diversion of human labor into agriculture if a catastrophic decline in food production is to be avoided. We have also seen (Chapter 11) that as energy dwindles, human beings must redistribute in accord with the natural carrying capacity of the earth. These logical consequences of energy and resource depletion imply a policy of large-scale resettlement of the human population, culminating in a predominantly rural civilization. As a means toward this end, one can imagine the evolution of urban farms that both provide employment and also serve as a training ground for the resettlement of city dwellers to self-sufficient rural communes based on labor-intensive agriculture. The possibilities for imaginative integration of urban and farm policies are endless. The use of the military services as a source of domestic labor, as practiced in the USSR and China, could facilitate the immense transition. Only the future will tell whether a democratic society can implement in advance long-term urban policies that flow with, rather than against, the tide of history.

NOTES

1. K. Davis, "The Urbanization of the Human Population," *Scientific American*, 213 (1965): 40–52.

2. V. P. Barabba, "The National Setting: Regional Shifts, Metropolitan Decline and Urban Decay," in *Post-Industrial America: Metropolitan Decline and Inter-Regional Job Shifts*, edited by G. Sternlieb and J. W. Hughes (New Brunswick, N.J.: Rutgers University Press, 1975), pp. 39–76.

3. Ibid.

4. P. A. Morrison, "Urban Growth and Decline: San José and St. Louis in the 1960's," *Science* 185 (1974): 757–62.

5. Ibid.

6. Barabba, "The National Setting."

7. Ibid.

8. "Poll Reports a Third of People in Cities Want to Move Out," *New York Times*, 2 March 1978, p. A14.

9. M. R. Greenberg, and N. Valente, "Recent Economic Trends in the Major Northeastern Metropolises," in *Post-Industrial America: Metropolitan Decline and Inter-Regional Job Shifts*, edited by G. Sternlieb and J. W. Hughes (New Brunswick, N.J.: Rutgers University Press, 1975), pp. 77–99.

10. G. Sternlieb and J. W. Hughes, "Is the New York Region the Prototype?" in *Post-Industrial America: Metropolitan Decline and Inter-Regional Job Shifts*, edited by G. Sternlieb and J. W. Hughes, (Eds.), (New Brunswick, N.J.: Rutgers University Press, 1975), pp. 101–37.

11. Ibid.

12. Ibid.

13. S. Rattner, "16% Rise Last Winter Makes New York Gas Costliest," *New York Times*, 9 July 1977, p. C23.

14. T. Muller, "The Declining and Growing Metropolis—A Fiscal Comparison," in *Post-Industrial America: Metropolitan Decline and Inter-Regional Job Shifts*, edited by G. Sternlieb and J. W. Hughes (New Brunswick, N.J.: Rutgers University Press, 1975), pp. 197–220.

15. Ibid.

16. P. Kihss, "Taxes Found Crucial in Rise in Living Costs," *New York Times*, July 1977, p. 1.

17. I. Hoch, "City Size, Effects, Trends and Policies," *Science* 193 (1976): 856–63.

18. P. Delaney, "Many Cities Like New York, Making Do with Less," *New York Times*, 31 March 1977, p. 1; H. Giniger, "Montreal Slogging Through Winter of Unusual Discontent," *New York Times*, 14 January 1978, p. 2; J. T. Wooten, "Aging Process Catches up with Cities of the North," *New York Times*, 13 February 1976, p. 1.

19. Hoch, "City Size, Effects, Trends and Policies.

20. For elaboration, see Muller, "The Declining and Growing Metropolis," p. 208.

21. "The Mayor's Executive Budget," *New York Times,* 15 April 1976, p. 23.

22. M. Friedman, writing in *Newsweek,* 17 November 1975, p. 90.

23. T. N. Clark, I. S. Rubin, L. C. Pettler, and E. Zimmerman, "How Many New Yorks? The New York Fiscal Crisis in Comparative Perspective," Research Report #72 of the Comparative Study of Community Decision Making (available from the first author, Department of Sociology, University of Chicago, Chicago, Ill. 60637).

24. U.S. Bureau of Labor Statistics, cited in Clark et al., "How Many New Yorks?"

25. P. Kihss, "Study Shows 7 Cities Have More Ineligibles on Welfare than New York," *New York Times,* 13 December 1977, p. 36.

26. Clark et al., "How Many New Yorks?"

27. J. T. Wooten, "Philadelphia Also Facing Crisis of Fund Shortage," *New York Times,* 20 May 1976, p. 1.

28. W. Serrin, "The Decline and Fall of Detroit," *New York Times,* 25 August 1976, p. 33; R. Wilkins, "The Motor-City Blues," *New York Times,* 20 August 1976, p. A21.

29. W. K. Stevens, "Cleveland Schools Face Shutdown on Friday for a Lack of Money," *New York Times,* 17 October 1977, p. 1; R. A. Stuart, "Cleveland Schools in Financial Limbo," *New York Times,* 8 April 1978, p. 9.

30. W. King, "Atlanta's Upward Surge Falters as City's 2nd Hotel Complex Fails," *New York Times,* 11 February 1978, p. 1; J. P. Sterba, "Houston Tangles with the Problems of Success," *New York Times,* 16 December 1977, p. B1; G. Lichtenstein, "Modern Problems Add Harsh Tones to Palette of Artistic Old Santa Fe," *New York Times,* 13 May 1976, p. 33.

31. H. Giniger, "Montreal Slogging Through Winter of Unusual Discontent," *New York Times,* 14 January 1978, p. 2.

32. P. T. Kilborn, "Glasgow, the City of Slums, Is Defying Rehabilitation," *New York Times,* 14 April 1976, p. 14.

33. R. Reed, "Britain, in a Reversal, Tries to Stem London Exodus," *New York Times,* 30 June 1977, p. 1.

34. B. Ward, *The Home of Man* (New York: W. W. Norton, 1976), p. 105.

35. A. H. Malcolm, "In Japan, Too, They Dream of a Little House with a Garden," *New York Times,* 3 March 1976, p. 31.

36. Id., "Japanese Cities Running Deficits," *New York Times,* 19 February 1976, p. 13; id., "Tokyo, on the Brink of Bankruptcy, Facing Central Government Rule," *New York Times,* 4 February 1978, p. 1.

37. D. K. Shipler, "Ideal Planning an Elusive Goal in Soviet, Too," *New York Times,* 10 June 1977, p. B1.

38. W. Thompson, "Economic Processes and Employment Problems in Declining Metroplitan Areas," *Post-Industrial America: Metropolitan Decline and Inter-Regional Job Shifts,* edited by G. Sternlieb and J. W. Hughes (New Brunswick, N.J.: Rutgers University Press, 1975), pp. 187–96.

39. W. Peterson, *Population,* 3rd ed., (New York: Macmillan, 1975), p. 478.

40. "Rural Areas, Small Towns, Grow Faster than Cities," *New York Times*, 2 February 1977, p. 8.
41. Barabba, "The National Setting."
42. Ibid.
43. Malcolm, "In Japan, Too . . ."
44. "Rural Areas, Small Towns, Grow Faster Than Cities."
45. M. Sterns, "Union Carbide, 3,500 on Staff, to Quit City," *New York Times*, 20 March 1976, p. 1; "Texasgulf's Moving out of the City to Connecticut," *New York Times*, 11 March 1976, p. 19; T. Robards, "Marine Midland Planning Layoff," *New York Times*, 24 March 1976, p. 47; M. Sterne, "The Hopes for Reviving City's Ailing Economy Lies in Easing Disadvantages of Doing Business Here," *New York Times*, 11 March 1976, p. 25.
46. J. P. Fried, "Housing Construction Below 1932 Level Here," *New York Times*, 15 March 1976, p. 53; I. Barmash, "Sales at City's Big Stores down 7.6% in May, the Sharpest Drop Since 1966," *New York Times*, 3 June 1976, p. 55; C. B. Horsley, "Skyscraper Faces a Bank Takeover," *New York Times*, 12 March 1976, p. 36.
47. M. Sterne, "Joblessness Is up to 12.2% in City," *New York Times*, 3 March 1976, p. 1; F. X. Clines, "Beame Proposes New Budget Cuts Ending 8,000 Jobs," *New York Times*, 2 March 1976, p. 1; S. R. Weisman, "Layoffs of 2,000 Reported a Part of State Budget," *New York Times*, 15 March 1976, p. 1; M. H. Seigel, "State Court Upholds Layoffs by Cities," *New York Times*, 27 January 1976, p. 27.
48. F. X. Clines, "A Shortfall of $500 Million Is Feared in Beame's Plan," *New York Times*, 26 April 1976, p. 1; id., "Revision Expected to Put City Deficit Near a Billion," *New York Times*, 12 February 1976, p. 1; M. Tolchin, "Doubt Expressed on City Finances," *New York Times*, 3 April 1976, p. 1; E. Ranzal, "State Finds Deficit Is Greater than New York City Expects," *New York Times*, 22 May 1976, p. 1; L. Dembart, "Secret New York Draft Predicts Increasing Budget Deficits for City," *New York Times*, 15 December 1977, p. 34; id., "Rohatyn Warns City of New Fiscal Crisis," *New York Times*, 12 January 1978, p. 1.
49. F. Ferretti, "City Seen Entering Era of Retrenchment," *New York Times*, 2 February 1976, p. 1; M. Knight, "Connecticut's Bond Rating Reduced 2nd Time in Year," *New York Times*, 6 March 1976, p. 42; F. X. Clines, "Problems Seen by M.A.C. in New York Cash Flow," *New York Times*, 3 June 1976, p. 32; S. R. Weisman, "Low Moody's Rating Compels New York to Cancel Note Sale," *New York Times*, 11 November 1977, p. 1.
50. M. Tolchin, "Senate Votes Shift in Funds for Housing to the Older Cities," *New York Times*, 8 June 1977, p. 1; E. C. Burks, "Carter Acts to Back $2 Billion in Bonds for New York City," *New York Times*, 3 March 1978, p. 1; R. Reinhold, "H.U.D. in 'Major Policy Change,' Increasing Aid to Central Cities," *New York Times*, 28 December 1977, p. 14.
51. F. X. Clines, "U.S. Is Seeking to Tighten Fiscal Monitoring of City," *New*

York Times, 20 January 1976, p. 15; I. Peterson, "U.S. Checks State Budget to Be Sure It's Balanced," *New York Times*, 17 February 1976, p. 1.

52. F. X. Clines, "Control Board Votes to Cut City Paychecks $6 a Week," *New York Times*, 28 February 1976, p. 1; L. Dembart, "Panel Bids Beame Cut Benefits Paid to City's Workers," *New York Times*, 3 June 1976, p. 1; S. R. Weisman, "Beame Is Preparing Drastic Cutbacks in City's Spending," *New York Times*, 11 October 1976, p. 1; L. Dembart, "Koch Will Impose Freeze on Hirings, Cut Appropriations," *New York Times*, 17 January 1978, p. 1; D. Stetson, "New York City Asks a Rollback in Employee Benefits," *New York Times*, 28 February 1978, p. 1.

53. J. Feron, "Garbage Clean-Up Begins in Yonkers," *New York Times*, 27 January 1976, p. 27; J. P. Fried, "East Side Tenants Lose Heat," *New York Times*, 5 February 1976, p. 35; E. C. Burks, "Jersey Bus Service Cutbacks Jolt Riders," *New York Times*, 14 February 1976, p. 29; F. J. Prial, "Jersey Bus Strike Disrupts 450,000," *New York Times*, 10 March 1976, p. 1; D. Stetson, "Health Cut Bills Stir Strike Talks," *New York Times*, 11 March 1976, p. 42; id., "Apartment Workers Strike at Many Buildings," *New York Times*, 4 May 1976, p. 1; id., "Owners Demand Pickup of Refuse," *New York Times*, 7 May 1976, p. 1.

54. F. X. Clines, "Curb Is Imposed on Transit Pact by Fiscal Board," *New York Times*, 19 May 1976, p. 1; S. R. Weisman, "Carey's 'New Era' Budget of $10.76 Billion Provides Deep Cuts, No New Taxes," *New York Times*, 21 January 1976, p. 1.

55. M. Sterns, "State's Commerce Chief Asks Fiscal Shift to Right," *New York Times*, 2 February 1976, p. 1; L. Greenhouse, "Carey Tilts to Economy at Expense of Environment," *New York Times*, 6 February 1976, p. 26; I. Peterson and R. Severo, "Carey Bids the Legislature Delay Environmental Law," *New York Times*, 17 February 1976, p. 1; L. Fellows, "Grasso Budget Would End Tax on Business Services," *New York Times*, 5 February 1976, p. 1.

56. S. Raab, "Major Crime up 11.8% Here in '74," *New York Times*, 10 February 1976, p. 1; id., "Crime Rose in Richer Neighborhoods, Fell in Poorer Sections of City in 1975," *New York Times*, 15 March 1976, p. 1; id., "Felony Arrests Rise 6% in New York City," *New York Times*, 21 September 1976, p. 1; M. Knight, "Vandalism Puzzles Suburbs," *New York Times*, 22 April 1976, p. 35; P. Delaney, "Suburbs Fighting Back as Crime Rises," *New York Times*, 30 August 1976, p. 1; P. Gupte, "Calls Swamp Police 911 Emergency Line," *New York Times*, 19 August 1976, p. 1; R. Stuart, "Detroit Recalling Police in Crime Wave," *New York Times*, 17 August 1976, p. 1.

57. R. Wilkins, "Surgery or Suicide," *New York Times*, 30 March 1976, p. 31; "Distress Signal," *New York Times*, 2 February 1976, p. 22.

58. "Violence of Youth Gangs Is Found at a New High," *New York Times*, 1 May 1976, p. 21; A. Salpukas, "Vicious Youth Gangs Plague Detroit," *New York Times*, 18 August 1976, p. 1.

59. "Detroit Is Tightening Its Curfew in Effort to Combat Youth Gangs," *New*

York Times, 21 August 1976, p. 19; A. Salpukas, "Detroit Police Chief Told to Stop Gangs," *New York Times*, 19 August 1976, p. 60.

60. G. Vecsey, "For Young Urban Nomads, Home Is in the Street," *New York Times*, 1 June 1976, p. 33.

61. N. Sheppard, "More Teen-aged Girls Are Turning to Prostitution as Means of Survival," *New York Times*, 4 May 1976, p. 39.

62. M. Waldron, "Jersey Foresees Crisis in Benefits for the Unemployed," *New York Times*, 21 January 1976, p. 37.

63. G. Goodman, "New York City Budget Cuts Said to Hurt Children," *New York Times*, 3 May 1976, p. 21; P. Kihss, "City Urged to Cut Transit, College and Hospital Aid," *New York Times*, 8 March 1976, p. 1.

64. F. X. Clines, "City Layoffs Hurt Minorities Most," *New York Times*, 20 February 1976, p. 1; Goodman, "New York City Budget Cuts Said to Hurt Children."

65. I. Peterson, "State Acts to Cut Day-Care Center Costs," *New York Times*, 6 February 1976, p. 26; P. Kihss, "New York City to Halt Aid for 49 Day-Care Centers," *New York Times*, 28 May 1976, p. 1.

66. M. Breasted, "Council and Students Fight to Save Life of Adult Classes," *New York Times*, 21 January 1976, p. 33.

67. F. X. Clines, "City U. Is Closed by Kibbee Pending a Fiscal Solution," *New York Times*, 29 May 1976, p. 1; id., "City University Unable to Meet Its Faculty Payroll Due Today," *New York Times*, 28 May 1976, p. 1; J. Cummings, "City U.'s Freshman Class to Shrink 40% Next Fall," *New York Times*, 20 April 1976, p. 1; F. X. Clines, "City to End Its Aid to 4-Year Schools at City U. in 1977," *New York Times*, 13 March 1976, p. 1; L. Buder, "Schools Maintenance Chief Urges Custodial Cutbacks," *New York Times*, 27 January 1976, p. 33.

68. N. Sheppard, "New York to Cut Summer Job Plan," *New York Times*, 22 April 1976, p. 19.

69. H. Faber, "New York State Is Closing Its Parks One Day a Week," *New York Times*, 17 April 1976, p. 1.

70. F. Cerra, "Budget Cuts Curbing Consumer Agency," *New York Times*, 24 April 1976, p. 1.

71. M. Schumach, "Carnegie Hall to End Its Live-in Studios for Artists," *New York Times*, 14 November 1977, p. 1.

72. "City, Saving $200,000, Drops 12 Jobs in Health Services Unit," *New York Times*, 21 January 1976, p. 39; I. Peterson, "Local Care Urged for Mentally Ill," *New York Times*, 2 March 1976, p. 1; id., "Carey Calls for Local Care of the Retarded and Unruly," *New York Times*, 21 January 1976, p. 39; "Director of Hospital Fund Says 2,800 Beds Must Go in the City," *New York Times*, 4 February 1976, p. 17; D. Bird, "Cut in Hospitals Called for Here," *New York Times*, 21 January 1976, p. 1; L. Maitland, "City Health Agency Votes to Shut 4 Hospitals in Plan to Cut Budget," *New York Times*, 6 February 1976, p. 26; D. Bird, "City Decision to Shut 4 Hospitals approved by State Health Chief,"

New York Times, 13 March 1976, p. 12; id., "State Study Asks City to Dissolve Hospital Agency," *New York Times*, 21 February 1976, p. 1.

73. R. Sullivan, "A Virtual Lack of Doctors Found in Some Slum Areas of New York," *New York Times*, 13 December 1977, p. 45; M. Sterne, "Residents of Harlem Suffer Worst Health in New York," *New York Times*, 10 April 1978, p. 1; "Studies Show That Poor Health Correlates with Acute Poverty," *New York Times*, 10 April 1978, p. D8; R. Sullivan, "U.S. to Send 20 to 50 Doctors to Poorer Sections of New York," *New York Times*, 31 January 1978, p. 1.

74. Muller, "The Declining and Growing Metropolis"; M. M. Webber, "San Francisco Area Rapid Transit—A 'Disappointing' Model," *New York Times*, 13 November 1976, p. 23.

75. G. Fowler, "Doubling Transit Use in Cities Found Unlikely to Lessen Autos," *New York Times*, 23 August 1976, p. 1.

76. D. Stetson, "Drop in Subway Use Laid to Economy," *New York Times*, 26 March 1976, p. 60.

77. E. C. Burks, "215 More Daily Subway Runs Will Be Eliminated by Aug. 30," *New York Times*, 14 August 1976, p. 1; E. C. Rurks, "New York Transit System Facing Necessity for Further Cutbacks," *New York Times*, 7 December 1976, p. 1; W. Saxon, "$30 Million Cut Proposed to Meet New York Transit Agency Shortage," *New York Times*, 15 December 1976, p. 1.

78. See above, note 53; F. J. Prial, "Tenants in Five Buildings without Heat for a Month," *New York Times*, 31 January 1976, p. 20; J. P. Fried, "East Side Tenants Lose Heat," *New York Times*, 5 February 1976, p. 35.

79. A. S. Oser, "A South Bronx Landlord Says Poverty Is Not to Blame for Blight," *New York Times*, 14 October 1977, p. B3.

80. J. P. Fried, "City's Rent Rises Outpace Tenant Income Increases," *New York Times*, 19 January 1976, p. 1.

81. Id., "Housing Abandonment Spreads in Bronx and Parts of Brooklyn," *New York Times*, 12 April 1976, p. 1.

82. Id., "Ravaged Sentinels of Blight, Abandoned Buildings," *New York Times*, 21 August 1976, p. 25.

83. C. Kaiser, "Total Realty Value Shows a Drop Here, First in 33 Years," *New York Times*, 3 February 1976, p. 1.

84. G. Sternlieb and J. W. Hughes, Introduction to Part II, *Post-Industrial America: Metropolitan Decline and Inter-Regional Job Shifts*, edited by G. Sternlief and J. W. Hughes (New Brunswick, N.J.: Rutgers University Press, 1975), pp. 169–70.

85. E. M. Hoover and R. Vernon, *Anatomy of a Metropolis* (Cambridge, Mass.: Harvard University Press, 1959), p. 57.

86. Thompson, "Economic Processes and Employment Problems . . . ," p. 194.

87. Greenberg and Valente, "Recent Economic Trends . . . ," p. 97.

88. Muller, "The Declining and Growing Metropolis," p. 199.

89. Ibid., p. 216.

90. J. F. Kain, "Implications of Declining Metropolitan Population on Hous-

ing Markets," in *Post-Industrial America: Metropolitan Decline and Inter-Regional Job Shifts*, edited by G. Sternlich and J. W. Hughes (New Brunswick, N.J.: Rutgers University Press, 1975), pp. 221–27 (quoted material from p. 227).

91. G. Sternlieb and J. W. Hughes, Prologue to *Post-Industrial America: Metropolitan Decline and Inter-Regional Job Shifts*, edited by G. Sternlieb and J. W. Hughes (New Brunswick, N.J.: Rutgers University Press, 1975), pp. 1–25 (quoted material from p. 17).

92. Kain, "Implications of Declining Metropolitan Population," p. 23.

93. Sternlieb and Hughes, "Is the New York Region the Prototype?" p. 136.

94. R. L. Meier, "A Stable Urban Ecosystem," *Science* 192 (1976): 962–67 (quoted material from p. 962).

95. D. E. Rosenbaum, "Six Cities Win Federal Grants to Test Teen-age Job Program," *New York Times*, 11 January 1978, p. 1.

96. M. Schumach, "100 'Farms' Planned on Lots in New York," *New York Times*, 26 May 1976, p. 1.

Chapter 13
VISIONS

Throughout this book I have stressed that nature is cyclic in its ways. Water, we have seen, is propelled by the sun from sea to sky and back again to sea. The elements in our bodies are borrowed from the earth, to be returned when life departs and used again by future generations. The surf, the tides, the moons, the seasons, the climate—all fluctuate on a regular rhythm ranging from seconds to 100,000 years. The universe itself may expand and collapse on a regular schedule of billions of years, in an endless cosmic cycle of death and rebirth.

Human civilizations likewise rise and fall, perhaps subject to the same pervasive law of nature. Historians have noted the cyclic character of past civilizations,[1] but convincing explanations of the underlying causes have proved elusive. The example of our own society, however, suggests a potentially general model. As a civilization grows by positive feedback, it unavoidably consumes the physical resources on which it depends. When growth encounters the limits set by a finite resource base, negative feedback prevails and growth ends. By this stage the civilization has unavoidably outgrown its resource base; its very maintenance demands a greater flow of resources than can be sustained by the dwindled supplies. Economic and physical contraction therefore ensue, amplified as we have seen by positive feedback. Thus, by this model, the history of human civilizations closely parallels the bell-shaped curve of resource depletion.

In this book we have in essence applied this hypothetical model to our own civilization (Figure 13.1). We have seen how basic landmarks of our society might change with the depletion of energy and natural resources, including money and wealth (Chapter 5), prevalent occupations (Chapters 8 and 9), and the size of the human population (Chapter 11). In this chapter we will extend the same reasoning to other aspects of life in an energy-scarce future. We will inquire how the decline of energy and resources might alter habitats, work, and education, and we will speculate on the future of political institutions, culture, and values. In discussing these issues we will continue to sketch in the imagination the broad outlines of a civilization that may replace our own; and we will see that in comparison with the old, the new civilization need not be uninviting.

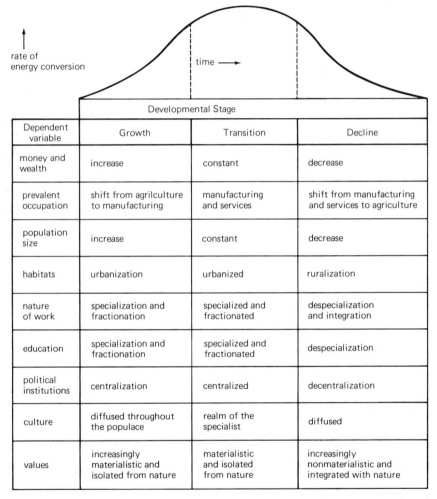

Developmental Stage			
Dependent variable	Growth	Transition	Decline
money and wealth	increase	constant	decrease
prevalent occupation	shift from agrilculture to manufacturing	manufacturing and services	shift from manufacturing and services to agriculture
population size	increase	constant	decrease
habitats	urbanization	urbanized	ruralization
nature of work	specialization and fractionation	specialized and fractionated	despecialization and integration
education	specialization and fractionation	specialized and fractionated	despecialization
political institutions	centralization	centralized	decentralization
culture	diffused throughout the populace	realm of the specialist	diffused
values	increasingly materialistic and isolated from nature	materialistic and isolated from nature	increasingly nonmaterialistic and integrated with nature

13.1   Life cycle of an industrial civilization. According to this hypothetical model, the major landmarks of industrial civilization (dependent variables) are shaped largely by the rate of energy conversion (independent variable). The first three dependent variables are discussed in earlier chapters, while the last six are discussed later in this chapter. According to the thesis of this book, industrial civilization has reached the stage of transition.

Living Patterns
The Decline
of the Traditional Community

Plato considered participation in community as a fulfillment of human nature, a source of individual identity and legitimization.[2] As many writers have noted, however, fossil fuels have dissolved the bonds of traditional community.[3] The personalized, sociable interactions that typify life in the traditional community have been replaced in our culture by the quick, impersonal, "functional" interchanges of the supermarket, the bank, and the fast-food stand. The Industrial Age has spawned a fragmented society of autonomous nuclear families, each housed separately in the ubiquitous, redundant tract home of the American suburb. Unprecedented personal mobility discourages people from settling in one place and developing a sense of community. In the absence of traditional community, its many social and emotional roles have been thrust upon the family unit, which understandably cracks under the strain.[4]

What has caused the demise of traditional community? Surely there are many reasons, some rooted deep in human nature. But the underlying motive force is the ready availability of energy and natural resources. The material excess born of abundant energy and resources permits the endless duplication of basic living facilities that underlies suburbanization, reducing the need to share and leading to the "luxury" of personal isolation. Likewise, it is abundant energy and resources that have conferred geographical mobility. It can be argued that industrialism contributes even to the deterioration of the family, for example, by redirecting the energies of both parents away from the home and toward occupational and material pursuits.

There are undeniable benefits to the industrial way of life. It is physically comfortable, free from material want, and food is comparatively plentiful. In particular, the broad exposure provided by personal mobility and the media can impart a worldly perspective, in sharp contrst to the narrow parochialism of many traditional cultures of the past. But there are clear costs to our way of life as well. In human terms, life in the suburbs can be lonely, rootless, and barren of deep meaning.

Cities in the Sky

Are our grandchildren destined also for a life in suburbia? There are those who believe "the big city is inevitable," based on a sevenfold increase in global population and a tenfold increase in per capita energy conversion.[5] But they have not considered where the prerequisite energy and resources might be found, nor how our sorely tested planet might absorb the implicit 70-fold increase in pollution. Others envision magnificent floating cities, with overall population densities of 300 people per acre.[6] But they have not explained how the human carrying capacity of a seagoing city can be elevated to more than 16 times the maximum realizable carrying capacity of arable land.

Perhaps the most ambitious vision of the future habitat is the space city—a self-sufficient colony several kilometers in diameter and housing up to 100,000 people[7] (Figure 13.2). According to their architects, such space cities would be assembled in stationary orbit above the earth, using resources hurled from the moon by gigantic interplantetary slingshots ("mass drivers"). Once erected, space cities would capture solar radiation, convert it to microwave radiation, and beam it to gigantic receiving stations on the earth. Thus space cities could in principle not only furnish a new human habitat, but also help relieve the shortage of energy and resources on the surface of the planet.

13.2 Artist's conception of a segment of a wheel-shaped space colony during the final stages of its construction from lunar materials. Note the agricultural areas, lakes, and rivers, which would be interspersed between densely populated areas. (from G. K. O'Neil, "Space Colonies: The High Frontier," *The Futurist* 10 [1976]: 25–33; courtesy NASA)

Space cities represent a well-meaning attempt to benefit industrialism and expand human horizons beyond our own planet. Moreover,
the U.S. Space Agency sees budgetary salvation in space cities and pro-
motes them with an enthusiasm that borders on fervor. It is therefore essential to look beneath the surface of these proposals, to inquire whether
they are realizable dreams and if so, whether they will truly benefit humanity. Proponents of space cities suggest that they will return their
enormous investment costs by opening new energy sources. In fact it took
10 years and $10 billion to put a human being on the moon. To build
mines, refineries, and factories on the moon would require hundreds—
and perhaps thousands—of billions of dollars, at a time when society is
hard pressed to find capital enough to maintain its energy sources on
earth (Chapter 8). Whether space colonies could return their gigantic investment costs depends in part on how long they would take to erect.
Their designers believe that the first space colony could be completed in
six years, the second in only two.[8] In fact it takes 10–12 years to build a
nuclear power plant on the surface of our planet. Perhaps the most telling
technical problem with space cities is biological in nature. In order to be
self-sufficient as planned, a space colony would have to be established as a
fully integrated ecosystem, from soil microorganisms to the plants
required for a breatheable atmosphere. In fact we lack the biological
knowledge to accomplish such a feat of creation even under controlled
conditions on the surface of our planet, let alone in the inhospitable environment of space. Above all, the architects of space cities have yet to
address the political realities and moral issues involved in spending what
may be the last of our treasures to construct gleaming space colonies for
an elite few while the cities on the surface of the planet collapse upon
their discontented and suffering masses.

At first sight proposals to convert the moon to space cities represent
an appealing dream, one that can easily grip the fertile human imagination. But space cities may also represent a new extreme in technological
escapism. Our limited resources may be better spent to improve the
human condition here at home, leaving the sky uncluttered and the moon
unspoiled for the gaze of future generations.

The Rebirth of Community

The earthly realities of declining energy and resources may spare our
grandchildren the burden of space colonies and the fate of the suburbs.
Cities will no doubt exist in an energy-scarce future, as they existed
before fossil fuels were harnessed, but the lack of energy and resources

may be expected to limit severely their size and number, and the great mass of human beings will presumably live outside their boundaries. The need to divert labor into agriculture (Chapter 10), the need to distribute people in accord with natural carrying capacities (Chapter 11), the inability to rebuild rapidly decaying cities (Chapter 12)—all of these forces are attributable ultimately to declining energy and resources, and all may be propelling the U.S. and the rest of the world toward a predominantly rural civilization.

We can only speculate on the organization of rural life in an energy-scarce future. *Homo sapiens* is a social animal; its natural inclination to live in groups, like the family, the tribe, and the village, may be coded indelibly in its genes. Moreover, the stringent labor requirements of self-sufficiency without abundant energy may encourage—and even compel—cooperation between individuals in the common pursuit of survival. Thus the decline of energy and resources might propel the evolution of cooperative, self-contained rural communities, in a process that Buckminster Fuller and others have termed the "retribalization" of industrial society.[9] The size of such rural communities will presumably be limited by the speed of prevalent nonmechanized modes of transportation and by local carrying capacities to a few hundred or thousand residents.

History is replete with examples of autonomous rural communities, from the titled estates of feudal Europe to the southern plantation of early U.S. history, the American Indian tribe, the Israeli kibbutzim, and Chinese agricultural communes. The "intentional" community of the contemporary U.S. provides a potentially relevant model for the future.[10] Several thousand such communities arose in the late sixties and early seventies,[11] largely as an idealistic, middle-class reaction to the values and lifestyle of modern industrial culture. For the present, such communities are bonded together less by material necessity than by a common religious or spiritual orientation, a shared purpose of vision that instills unity and restores meaning to the lives of their members. Few if any of these communities are truly self-sufficient, but most see this as an eventual goal. An ultimate aim of intentional communities in the U.S. has been to form a cooperative "federation of communities stretching from one end of the country to the other."[12] Such a cooperative confederacy is intended to replace today's "centralized social and political system with a loose network of small, decentralized communities made viable by small energy sources and miniaturized industries."[13]

The evolution of such future communities does not imply a return to an "idyllic" past. Not only was the past not necessarily idyllic, comprising as it sometimes did mind-numbing agricultural labor in the pursuit of

bare subsistence; in addition, the knowledge, worldly exposure, and sophistication provided by the experience of industrialism may prevent a
return to the past even if it were desirable. The problems of the future
will be unprecedented in the human experience, and so must be the solutions. The intentional communities of the contemporary U.S. represent
but one example of a new solution. Today such communities stand on the
fringes of American society as bold social experiments, but in an energy-
scarce future they could comprise the mainstream.

The Revaluation of Work

As implied throughout this book, the nature of work is historically related
to the availability of energy and natural resources. Prior to the Industrial
Revolution most people were farmers, as required by the unavailability
of energy (Chapter 9). Farmers were of necessity rude practitioners of
other trades as well, from carpentry to beekeeping, and hence work was
necessarily varied and despecialized for most people. Manufacturing—such that it was—was organized into crafts or skilled trades, with
production typically arranged so that a highly skilled worker was involved
in many or all stages, from conception to the final product. The service industries, constrained by the lack of a solid manufacturing base, were restricted to the priesthood and a small number of teachers, doctors, min-
strels, and administrators.

We have seen how the rise of fossil fuels changed agriculture from a
decentralized, subsistence operation to a centralized and highly specialized agribusiness, culminating in the monocrop (Chapter 9). Both manufacturing and the service industries have experienced a similar fate at the
hands of fossil fuels. As industrialism gathered momentum, the machines
of production grew ever larger and their manufactured products more in-
tricate. No longer could a single worker manage the entire production
process, which was accordingly split into sequences of ever-simpler tasks.
Such fractionation reaches its zenith in the modern assembly line, where
each worker is responsible for a minute step in the final assembly of a
manufactured product, but has no other role in production. Because energy and natural resources have been cheap and abundant, work has
"become increasingly subdivided into petty operations that fail to sustain
the interest or engage the capacities of humans."[14] The service industries, which grew as a consequence of manufacturing, became likewise
fragmented and specialized as the knowledge generated by industrialism

grew beyond the bounds of a single human intellect. No longer, for example, does one train to become simply a doctor; rather, one becomes a gynecologist, a gerontologist, or a gastroenterologist.

The large-scale, specialized, and fractionated production process made possible by abundant energy and resources lies at the heart of our affluent, mobile, and knowledgeable lifestyle. But the same process has also robbed people of jobs and work of its meaning. In times past, ample work was available for young and old alike. But the abundant and therefore cheap energy stored in fossil fuels has rendered human labor expensive by comparison. Reflecting this economic reality, a variety of social mechanisms now conspire to keep the old, and especially the young, off the labor market. As recently as 1955, 70% of those men between 60 and 69 were employed, but by 1976 the availability of Social Security and pension plans had reduced the figure to less than 50%.[15] In 1900 nearly one-fifth of all children between 10 and 15 worked;[16] today children are purposely excluded from the labor market until the age of 16, and people under 24 comprise the bulk of the unemployed throughout the U.S. and Western Europe.[17] Social mechanisms that limit the labor force include work permits, extended education, job seniority, retirement plans such as Social Security, and the very concept of "retirement" itself. Studies suggest that for the many Americans who find jobs, work is often dull, repetitious, and unsatisfying, motivated more by the weekly paycheck than by intrinsic reward.[18]

Work in our industrialized society is not only unfulfilling for many, it is also dangerous. Hundreds upon thousands of workers have been killed or crippled by the machines of modern production.[19] Occupational carcinogenesis[20]—work-related cancer—provides but one illustration of the health hazards of the modern workplace. The World Health Organization of the U.N. estimates that between 75% and 85% of all cancers are caused by environmental exposures,[21] many of which occur on the job. Of one million current and former workers in the U.S. asbestos industry, for example, an estimated 30%–40% can expect to die from work-related cancer.[22] In the steel industry, inhalation of ash from coal combustion has resulted in a lung cancer rate 10 times that of other steel workers.[23] A recent U.S. government report has found that, nationwide, one worker in four is exposed to hazardous conditions or substances whose long-range effects are only now becoming apparent.[24]

What has caused the degradation of work in industrial civilization? For Karl Marx, the culprit was capitalism. The fractionation of production into a sequence of unskilled tasks is seen by Marxists as a deliberate scheme to de-skill labor and thereby to wrest the control of production away from the working class. The scheme is effected, according to Marxist doctrine, by separating the planning of production from its execution. Planning is invested in an elite class of managers, the accomplices of the capitalist, in order to reduce the great mass of workers to replaceable cogs in the machine of production. By fragmenting production into a sequence of trivial tasks, unskilled and therefore cheaper labor can be employed, increasing profits to the capitalist, allowing faster accumulation of capital, and permitting greater capitalist control over the work force.[25]

Marx formulated his doctrines in early industrial England, where unbridled capitalism ruled the working class with a callous hand. In such a context it is not surprising that he attributed the social ills of the workplace to capitalism. Marxists still insist that the modern production process can be made humane by vesting full control of the means of production in the hands of the worker. They blame not the machine but the social organization that regulates its use. In fact, Marxism shares with capitalism a fundamentally similar viewpoint of human beings striving continually to "master" nature for their benefit.[26] Moreover, the history of industrialism suggests that communism and capitalism lead to similar modes of production. Centralized factories and assembly lines are the mainstay of manufacturing not only in capitalist countries but in Communist nations as well. Even the Chinese have acquired the attitude that nature is to be "mastered."[27] According to its present leadership, China now seeks rapid economic development and a technological revolution, incorporating material incentives for workers.[28] Likewise, manufacturing in the USSR is organized much like its American counterpart; the principal difference is that Soviet workers "own" the means of production through the state, while U.S. workers "own" them through the stock market. In capitalist and socialist states alike, manufacturers are driven to maximize returns to permit reinvestment and therefore the positive-feedback accumulation of capital that underlies industrialism. Even pollution, the hallmark of industrialism, is nearly as rampant in the USSR as in the U.S.[29] As other authors have noted,[30] differences in political or economic philosophy are overshadowed by the structural similarities inherent in the industrial process itself. Seen in this light, specialization and fractionation of production may be less the consequences of capitalism than inevitable human responses to complexity, intrinsic marks of the modern machine under any socioeconomic system.

According to the above reasoning, it is the large-scale use of energy and natural resources that shapes the character of work in our industrial civilization. The decline of energy and resources would thus imply a fundamental restructuring of work. We have seen that if energy and resources become scarce and more expensive, human labor will have to be diverted from services and manufacturing to agriculture if widespread starvation is to be avoided (Chapter 10). Here may lie the key to the nature of work in an energy-scarce future. If our central assumption—the decline of energy and resources—is validated as the future unfolds, then agriculture must be the primary occupation of any future civilization. As in times past, the bulk of humanity may labor in the fields to produce food.

Now, as in the past, farming is an inherently diverse undertaking. It requires the skill of the mechanic, the knowledge of the veterinarian, an understanding of countless plant and animal species, and a feel for the soil and the seasons. Thus the average inhabitant of tomorrow's rural community may be skilled in many trades and required by circumstance to undertake tasks that are performed today by a variety of specialists. With the reduction of large-scale manufacturing, centralization and fractionation of the production process may be expected also to decline. The crafts and skilled trades may reassert themselves as the dominant means of production, and with them may come the satisfaction of involvement in all stages of the work process. Leisure may become more limited, and "unemployment" a distant memory of a time when fossil fuels temporarily displaced human labor. Young and old alike may be reintegrated into work of necessity as communities harness all available labor in the quest for self-sufficiency. The harsh labor requirements of the self-sufficient lifestyle might encourage the willing use of whatever laborsaving technologies are available, from the tools of agriculture to decentralized, small-scale power generators driven by the sun and wind. With the decline of money (Chapter 5), the direct exchange of foods and services—barter—may necessarily return to American life. Indeed, in some quarters it already has.[31] In the absence of abundant energy and resources, life in the rural community of the future may be demanding; but it may also be balanced and rewarding, with value and meaning restored to even the most menial chore. Work in America may be redefined, from today's monotonous pursuit of paycheck to "the material expression of love."[32]

The Restructuring of Schooling

It is said that the character of a society can be deduced in large part from the way it organizes work. The structure of educational institutions in par-

ticular derives from the structure of work, since schooling is the primary
institutional means by which young minds are shaped to serve the existing socioeconomic system. As observed by Bowles and Gintis,

> the structure of the educational experience is admirably suited to nur-
> turing attitudes and behaviors consonant with participation in the labor
> force. Particularly dramatic is the statistically verifiable congruence be-
> tween personality traits conducive to proper work performance on the
> job and those which are rewarded by high grades in the classroom.[33]

The parallels between the classroom and the workplace experiences
are profound. Both in the schoolroom and on the job respect for authority
is essential, punctuality is rewarded, absenteeism is discouraged by pen-
alty, and schedules are rigidly locked to the clock. The school bell furnishes early conditioning for the factory whistle in later life, as young
minds are trained in the concept of time that is so essential to effective
subsequent participation in the labor force. In the classroom students are
conditioned to look outside themselves for reinforcement by the practice
of grading, in preparation for similar externalization of reward later in the
workplace. The class divisions between planner-managers and workers
which characterize the workplace appear early in the educational experience as children are segregated into groups on the basis of academic
achievement and their conformity to the expectations of the classroom.
Fractionation and specialization, also hallmarks of the work experience,
appear early in the classroom as knowledge is compartmentalized into different "subjects." The higher one progresses in education, the narrower
and more specialized it becomes. By the time students reach college, the
educational system insists that they focus the bulk of their attention on a
single fragment of human knowledge, to the exclusion of all others.

And then, as suddenly as it began, the educational experience ends.
Young adults are thrust into the workplace to take their turn pulling the
levers of industrialism. The low-achievers are abandoned first by the educational system, prevented by lower scholastic marks from continuing
beyond secondary school. From their ranks industrial society has traditionally drawn its janitors, assembly-line workers, and post office personnel. High-achievers, in contrast, have traditionally been "promoted" to
the university, where they train to enter the meritocracy and become the
planner-managers of industrial society. Little value has been placed on
lifelong study unless its purpose is to sharpen and update job skills that
are needed for more effective work performance. Education has become
the province of the young in industrial society.

As work is restructured by the depletion of energy and resources, education may be expected to follow suit. Early signs of potentially fundamental change in the educational system are already visible, as highly specialized recipients of the Ph.D. degree are unable to find employment,[34] many teachers are jobless,[35] and the mere possession of a college degree no longer guarantees a high-paying job. Funds for university research are declining,[36] specialized graduate departments are contracting[37] and diversifying,[38] and applications to professional schools are ebbing.[39] Undergraduate enrollments too are declining at many American universities,[40] and will continue to contract for years to come owing to the age structure of the American population (Chapter 11). University budgets have in the past been linked with enrollments; if the linkage is maintained, then financial support for universities may be expected to decline. In times of economic contraction, even elementary schools must struggle to remain solvent and open.[41]

In an energy-scarce future, education may have a much different definition than in our present society. The historic role of education, preparing young minds for participation in society, will presumably continue; but as the structure of society changes, so may the emphasis of education. The decline of transportation, the decentralization of institutional life, and the curtailed financial support from state and federal sources may shift the responsibility for education to local initiative. The result may be greater diversity in the purposes and organization of the educational experience. As specialization declines in the workplace, it may decline also in the schoolroom. Study may be more closely integrated with work throughout life, as in contemporary China, where "Studying is sort of like eating. We do it all the time."[42] With most people involved in producing food, preparation for work may take the form of on-the-job experience and apprenticeships. Education may supplement such work preparation, but also become more a source of enjoyment and self-fulfillment. The decline of energy and resources may redefine education, and reverse its mission of specializing the human intellect and thereby narrowing horizons, to expanding human awareness of the material and spiritual universe.

The Decentralization of Political Institutions

Historically, the chief role played by governments has been the allocation of resources for various of society's purposes, including especially defense from external forces and the maintenance of internal social order. Hence as the availability of energy and resources has increased, the size, strength, complexity, and budgets of central governments have increased in almost exact proportion. Centralized bureaucracies now govern throughout the industrialized world, obtaining revenues largely by taxation and allocating the collected treasure for government, defense, and countless social programs that function to redistribute wealth internally.

Bureaucracies, however, are also apparently subject to the ubiquitous law of diminishing returns. As they become larger and more complex, bureaucracies may also become unwieldy and rigid, and increasingly incapable of effectively addressing problems they were intended to solve.[43] In an energy-scarce future, the revenues available to governments seem destined to decline, further limiting the capacity of government to "solve" the problems of society. Taxpayer revolts are evident already in some states, and more than half the American population distrusts government.[44] As wealth declines this revolt of the middle class may intensify, affecting all aspects of government spending. For example, industrial societies may be increasingly unable to afford massive military establishments on today's scale. Military budgets may contract, and new emphasis may be placed on the defense of home territory as opposed to offensive capacity abroad. Likewise, central government spending on its own operations, and on various internal social programs, may shrink as energy and resources are depleted.

The character of government in an energy-scarce future is difficult if not impossible to foresee. Heilbroner anticipates the collapse of democratic governments and the rise of centralized, authoritarian regimes as the world's mounting problems require swifter and more radical remedies.[45] It is debatable whether such a turn of events would represent a significant global change; a recent study suggests that even today, only 19.6% of the world's population lives in "free" nations[46] (Figure 13.3). In the early stages of the transition brought about by energy and resource scarcity, increased power of centralized government would seem a logical and precedented response to mounting socioeconomic problems. Such power could be exercised despite shrinking budgets through governmental control of information (the press and television networks) and by selective government allocation of scarce resources, as during past world wars and depressions.

On the other hand, the decentralization of society that we have discussed may compel the eventual decentralization of political power as well. The distribution of decision-making power is already evident in hard-pressed urban areas, as illustrated by New York's "Community

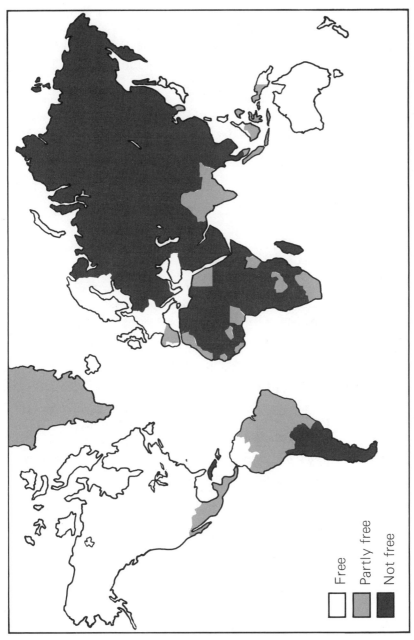

13.3 Distribution of "free" nations and territories of the world. The map was created by Freedom House in accordance with how free it judges a people to be to select leaders, voice opinions without fear, and exercise freedom of religion, occupation and movement. (from D. Carmody, "Moves at International Parleys in '76 Seen as Threatening Free Press," *New York Times*, 22 December 1976, p. 12)

Free

Partly free

Not free

Boards."[47] Indeed, in the recent past, local governments in the U.S. have grown faster than the federal government in terms of budget. In the period from 1955 to 1974, federal government purchases fell from 11.1% to 7.9% of the Gross National Product (GNP), while state and local government purchases rose from 7.7% to 13.5% of the GNP.[48] A trend toward decentralization is not confined to the U.S., nor even to industrial nations; rather, it is global in scope.[49] If continued, trends such as these could culminate eventually in a confederacy of small, relatively autonomous local governments, perhaps coordinated and represented internationally by central governments that are vastly reduced from today's unwieldy bureaucracies.

Culture, Values, and Ethics

Abundant energy and resources have not only shaped the character of work, education, and politics in our industrial society; they have also molded our culture and inculcated us, consciously or otherwise, with some of our deepest values.

Concepts of Time

Consider, for example, the impact of industrialism on our concept of time. Prior to the Industrial Revolution, intervals of time were measured largely in terms of the cycles of nature—the position of the sun in the day–night cycle, the lunar month, the rhythm of the seasons. But early in the Industrial Age the process of modernization demanded a new concept of time. As work became increasingly fractionated, so also did time. The streetcar schedule, the time clock, the factory whistle—all are ultimate reflections of the need for mass coordination of workers in an industrial system, but all have helped assure that as industrialism grew, "the dictatorship of the clock and the schedule became absolute."[50] Today it is a rare member of industrial society whose life is not regulated by a timepiece strapped to the wrist.

Specialization

The impact of industrialism on modern values extends far beyond the concept of time. Cultural and occupational specialization, for example, are considered natural according to modern values. It has not always been so; in "primitive" hunter/gatherer societies, for example, art and music were diffused throughout the culture. Each individual typically served as a repository of the culture's musical and artistic traditions, and all participated in their expression. In the present age of specialization, however, music, drama, dance, and art have been dissected from the masses and pigeonholed in the realm of the specialist. Folk culture has declined in importance, to be replaced by specialized musical groups and dance troupes. In industrial societies status is even conferred according to the degree of specialization: the specialized university professorship is accorded high status while "general labor" is held in comparatively low regard. To be sure, specialization has permitted new heights in artistic achievement and new depths in human knowledge, but it has also relegated the role of most people to spectator of culture rather than participant.

Materialism

The ready availability of energy and resources may even be responsible for that most maligned of modern values, materialism. With the rise of industrialism came new notions of human progress, tied to the growth of capital. The generation of investment capital, termed "profit" in capitalist nations, is central to economic growth in all industrial societies, communist and capitalist alike. But without an active mass market to consume the many products of industrialism, the growth of capital would cease. Society's values have followed the economic imperative, and industrial civilization accordingly places a premium on material consumption. Indeed, the acquisition of material goods far beyond actual survival needs has become a value in itself, as evidenced by the lofty status that is widely accorded ownership of material wealth. Materialism is now propagated by the primary conditioning instrument of modern industrial society, television. The distribution of television broadcasting stations precisely delineates the extent and boundaries of industrial civilization (Figure 13.4). Talented advertisers, thoroughly trained in the modern science of psychology, wage an endless campaign to stir new desires, to create new markets where there is no genuine need, and to indoctrinate our impressionable children with these same values. The advertisers succeed; in so doing they reinforce materialistic values and help to accelerate the depletion of remaining supplies of energy and resources.

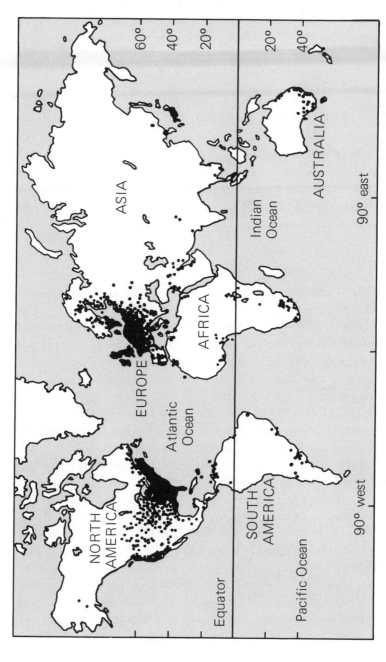

13.4 Distribution of world television broadcasting stations having an effective radiated power in excess of 50 kilowatts. Note the concentration in industrialized regions of the world, where television serves to propagate consumerism. (from the *New York Times*, 3 February 1978, p. A9)

The Decline of Diversity

Prior to the Industrial Age the peoples of the world evolved different customs, traditions, social mores, and values. Even within areas that today we call nations, the diversity of environments gave rise to a diversity of values, reflected in architecture, custom, and dress. But early in the process of industrialization factory life demanded regimentation of the labor force, and the common pursuit of material gain formed the basis of a monoculture that is global in scope. The rise of mechanized transportation and modern communication media, notably television and satellite transmission, and extensive economic interchange, have helped to homogenize remaining values within and between nations of the industrialized world. Coca-Cola and MacDonald's hamburgers are now marketed throughout the industrial world, and increasingly in nonindustrial nations. From Tokyo and St. Louis to Bonn and Moscow, cities now look alike, rapid growth is synonymous with good, individual conformity is valued, and the suit and tie comprise the universally accepted uniform of industrial civilization.

The Human Place in Nature

Nowhere is the effect of industrialism on values more profound than in the human image of self in relationship to nature. As we have noted, prior to the Industrial Age most people lived by farming. People were of necessity integrated into nature's cycles and attuned to the rhythm of the seasons. Humanity paid tribute to nature in drama, dance, and mythology, and deified natural forces in religion. Since the maturing of industrialism, however, most members of society have labored not in fields but in climate-controlled offices and factories. Abundant energy and resources have insulated the bulk of humankind from the powerful pulse of the seasons; giant reservoirs buffer us from flood and drought cycles; and electricity shields us from the darkness of the night. In a mere twinkling on the evolutionary time scale, the modest physical powers of the human species have been magnified supranaturally by abundant energy and resources. Cultural images embedded in our language betray the resulting attitude of dominion over nature. We "master" mountains, "conquer" seas, "harness" the atom, and even "tread upon" the virgin surface of the moon. Abundant energy and resources have at once uncoupled humankind from natural cycles and inculcated in industrial civilization the belief that human beings "control" nature. Certainly we affect nature with our activities, and accelerate many of her cycles through the use of fossil fuels.

But as we are periodically reminded by extreme weather and natural disasters, human control of nature is a myth.

The Birth of Eco-Logic

How might the decline of energy and resources compel our civilization to adopt new values that are more adaptive in an energy-scarce future? Without abundant energy and resources, transportation will slow and human beings must turn to agriculture for livelihood, as we have discussed. The rapid pace of life mandated by fossil fuels may slow, the rate of social change may slacken, and time may again be measured not by the streetcar schedule and the factory whistle, but rather by the rhythm of the moon and seasons. As energy and resources become scarce, society may no longer be able to afford the "luxury" of cultural specialization. We may witness a resurgence of folk culture as the practice of art, music, and drama becomes the province of the public rather than the specialty of a select few. The decline of transportation and the decentralization of institutions may allow natural variation in the environment to express itself in the diversification of human culture. Materialism may subside, its foundation eroded by the depletion of energy and resources. In the absence of material abundance the pursuit of gain may be supplanted by the pursuit of self-development, awareness, and actualization. The cultural preoccupation with the outer, materialistic world that unavoidably attends industrialism may be balanced by a growing concern for the inner, spiritual world.

Underlying and propelling all these cultural changes may be a redefinition of the human relationships with nature. If energy and resources decline in availability, it is unavoidable that human society will no longer be sheltered "artificially" from natural cycles. In an energy-scarce future, where the majority of people must grow, as well as bake, their daily bread, humankind may relearn the "wisdom of the soil." In the past, human understanding of its place in nature was intuitive, born of necessity in the struggle to survive, but never fully articulated in the human consciousness. The wisdom was grasped by the emotion but not the intellect, understood in the heart but not the mind. But now humanity is separated from its past by the immense body of knowledge accumulated during the Industrial Age. Fossil fuels have amplified the human senses with unprecedented technology and freed the restless human intellect to roam widely through the material universe. We understand in new light the biblical insight that our species arose from clay; the natural sciences have taught us that our very hearts and minds are made of elements in the soil, to which we return when the spirit departs. Ecology has revealed the intrinsic logic of karma and the Golden Rule: what we do unto nature, we do unto ourselves as well. We are in a position to understand that we do not *control* nature: we *are* nature, and it is us. There is no separation: all is one, propelled through different physical forms in space–time by nature's omnipresent cycles. The knowledge generated during the Industrial Age has paved the way for a historic fusion of the material and spiritual planes. The experience of industrialism has prepared humanity for the possibility of a new and enlightened relationship with nature, based not on contempt, dominion, and separation, but rather on respect, reciprocation, and integration. In an energy-scarce future we may be forced to incorporate this eco-logic into our culture in order to survive.

The Dawn Approaches

The Industrial Age has had massive impact on human consciousness, and its fruits still have a firm grip on the human imagination, especially in the nonindustrialized nations. Those countries that have an adequate domestic energy and resource base, such as China, the nations of Africa and South America, have made known their intent also to seek the benefits of industrialism. They will not be easily disuaded from passing through the same developmental cycle that the advanced industrial nations know so well. Thus the flames of industrialism may die to embers in one nation just as they are fanned in another, as the process spreads in successive waves through the cultures of our planet.

The advanced industrial nations, however, may have had their day in the sun. If the central assumption of this book is validated by future events—that is, if energy and natural resources become increasingly scarce and more expensive—then there will be no arresting the decline of the advanced industrialized nations. It has become customary to greet this possibility with pessimism. But in weighing any issue we are obliged to contemplate the alternatives—including in this case a future of unlimited energy and natural resources. If the past is an accurate indication, unlimited energy and resources would enable the human population to continue its two-century explosion, eventually overrunning every corner of the globe with human numbers. With unlimited energy and resources, most of the world's nations would presumably industrialize, obliterating the last vestiges of human diversity and creating a worldwide mono-culture. The machines of war would increase in numbers, sophistication, and destructive power. Air and water pollution would worsen manyfold; the seas might be poisoned beyond redemption; the fertile soils of the earth would be ravaged by industrial agriculture; and quiet forest retreats would become a distant memory. Eventually the unbridled industrial ac-tivity made possible by limitless energy and resources could precipiate frightful environmental calamities, such as drastic climatic changes in-duced by accumulated carbon dioxide. It is questionable how long this planet and its diverse life forms can bear the countless consequences of unlimited growth. It is debatable whether our species has demonstrated collective wisdom and compassion in its management of the supranatural powers conferred by abundant energy and resources. It is arguable that the costs of industrialism have begun to outweigh the benefits.

These considerations are not meant to denigrate the countless bene-fits of industrialism—the flourishing of the arts, the exhilarating advances in our comprehension of the universe, and the global perspective afforded by modern transport and the communication media. Indeed, the evolu-tion of a more enlightened culture would be unimaginable without this foundation. But the evidence summarized in this book suggests that we should make ready for change, and learn to see the best in it. If energy and natural resources have entered an era of decline, then optimism for the future cannot be based on the hope for a continuance of things past. Optimism for the future can be based on the expectation that the decline of one civilization promises the birth of another, more advanced civiliza-tion. Perhaps we stand at the threshold of a period of accelerated evolu-tion during which the forces of natural selection will favor those habits, traits, and values consistent with energy and resource scarcity. Our ca-pacity to learn and change may become the yardstick of human survival. The advanced industrial nations have been pushed to the leading edge of this developmental cycle, their collective consciousness raised and their sensitivity to the environment heightened by the experience of indus-trialism. History may have thrust upon these nations the challenge—and the opportunity—of building a new order.

The transition to a new order cannot be free of difficulty, but hard-ship is no stranger to our species. Among the greatest challenges of transi-tion is the redistribution of wealth within and between nations. As afflu-ence declines steadily, resistance to the redistribution of wealth may be expected to increase; but if social chaos is to be minimized, the need for such redistribution will also steadily increase. The response of society to this and like dilemmas may stamp the character of the approaching transition. Of course, we cannot know in advance whether humanity will handle the coming transition with grace and dignity, nor whether it will handle us gently. The human species is ingenious, adaptable, and tough; I do not doubt it will survive the birth of yet another civilization. If survival is the end, however, preparation is the means and now is the time. Industrial civilization may have seen its fullest flower, and material abundance may never surpass that of today.

The waning years of the 20th century represent an extraordinary period of human history. We may be witnessing no less than the grand finale of the Industrial Age, one of history's monumental cycles closing as yet another opens, and we are the privileged participants in this drama. As the drama unfolds, let us keep in mind that the cycles of nature and history imply continuous renewal. It is not the end, but the beginning. The turbulent events of the present are not death throes, they are the promise-laden pangs of birth. We witness not the final hour of the human species, but an epochal step in its continuing evolution. History now calls on us to bid the past farewell without regret, and to open our hearts and minds to the new.

NOTES

1. E.g., P. A. Sorokin, *Modern Historical and Social Philosophies* (New York: Dover, 1963), see especially pp. 275–322.
2. M. Bouvard, *The Intentional Community Movement: Building a New Moral World* (New York: National University Publications, Kennicat Press, 1975), p. 11.
3. E.g., ibid.; K. Melville, *Communes in the Counter Culture: Origins, Theories, Styles of Life* (New York: Morrow, 1972); B. Zablocki, *The Joyful Community* (Baltimore: Pelican, 1971).
4. P. Ariés, "The Family and the City," *Daedalus* 106 (1977): 227–37.
5. E.g., C. A. Doxiadis, *Anthropolis: City for Human Development* (New York: W. W. Norton, 1974), p. 25.
6. R. Mason, "Beyond 2000 Architecture," *The Futurist* 9 (1975): 235–246.
7. G. K. O'Neil, "Space Colonies and Energy Supply to the Earth," *Science* 190 (1975): 943–47; id., "Space Colonies: The High Frontier," *The Futurist* 10 (1976): 25–33.
8. O'Neil, "Space Colonies and Energy Supply to the Earth," p. 946.
9. H. R. Isaacs, *Idols of the Tribe* (New York: Harper and Row, 1975).
10. For background, see above, notes 2 and 3; R. Fairfield, *Communes USA, A Personal Tour* (Baltimore: Penguin, 1972); W. Hedgepeth, *The Alternative, Communal Life in America* (New York: Macmillan, 1970); R. M. Kanter, *Com-*

munes: Creating and Managing the Group Life (New York: Harper and Row, 1973); R. Roberts, *The New Communes: Coming Together in America* (Englewood Cliffs, N.J.: Prentice-Hall, 1971); R. Houriet, *Getting Back Together* (New York: Coward, McCann, Geoghegan, 1971).

11. Houriet, *Getting Back Together*.

12. Bouvard, *The Intentional Community Movement*, p. 125.

13. P. Goodman, quoted in Houriet, *Getting Back Together*, p. 6.

14. H. Braverman, *Labor and Monopoly Capital: The Degradation of Work in the Twentieth Century* (New York: Monthly Review, 1974), p. 4.

15. W. J. Eaton, "Workers 'Stampeding' to Early Retirement," *Los Angeles Times*, 12 June 1978, p. 1.

16. Braverman, *Labor and Monopoly Capital*.

17. J. Kandel, "Problem of Youth Joblessness Grips West Europe," *New York Times*, 4 February 1978, p. 25; C. H. Farnsworth, "Joblessness Among Youths Is Raising Worry in Europe," *New York Times*, 13 December 1976, p. 1; G. I. Maeroff, "Too Many Youths Found Aspiring to Too Few Jobs," *New York Times*, 9 November 1976, p. 18; A. Crittenden, "Finding a Solution to Unemployment," *New York Times*, 18 August 1976, p. 51; E. L. Dale, "I.M.F. Cautions Industrial Nations on Quick Bid to Cut Joblessness," *New York Times*, 20 September 1976, p. 49; J. Flint, "Rising Unemployment Bewilders Young Blacks," *New York Times*, 10 September 1977, p. 27; T. Wicker, "A Special Economic Problem," *New York Times*, 30 November 1976, p. 39C; R. Wilkins, "Surgery or Suicide," *New York Times*, 30 March 1976, p. 31.

18. S. Terkel, *Working* (New York: Avon, 1975); J. O'Toole, ed., *Work and the Quality of Life* (Cambridge, Mass.: MIT Press, 1974); H. G. Gutman, *Work, Culture and Society in Industrializing America* (New York: Knopf, 1976).

19. N. A. Ashford, *Crisis in the Work Place: Occupational Disease and Injury* (Cambridge, Mass.: MIT Press, 1976).

20. U. Saffiotti and J. K. Wagoner, *Occupational Carcinogenesis, Annals of the New York Academy of Science*, vol. 271 (New York: New York Academy of Science, 1976).

21. P. Lehmann, *Cancer and the Worker* (New York: New York Academy of Sciences, 1977), p. 3.

22. Ibid., p. 4.

23. Ibid.

24. D. Burnham, "U.S. Study Finds One in 4 Workers Exposed to Hazards," *New York Times*, 3 October 1977, p. 1.

25. E.g., Braverman, *Labor and Monopoly Capital*.

26. K. Marx, *Economic and Philosophical Manuscripts* (New York: Fredrick Ungar, 1961), p. 97.

27. D. H. Perkins, *Rural Small-Scale Industry in the People's Republic of China* (Berkeley and Los Angeles: University of California Press, 1977), p. 237.

28. F. Butterfield, "Peking Party Chief Pledges to Improve Standard of Living," *New York Times*, 27 February 1978, p. 1.

29. M. I. Goldman, "The Convergence of Environmental Disruption," *Science* 170 (1970): 37–42.
30. R. L. Heilbroner, *The Human Prospect* (New York: W. W. Norton, 1974).
31. R. Lindsey, "It's the Newest Trade-off: Bartering Your Services," *New York Times*, 18 November 1976, p. 48.
32. S. Gaskin, *Hey Beatnik! This Is the Farm Book* (Summertown, Tenn.: The Book Publishing Co., 1974).
33. S. Bowles and H. Gintis, *Schooling in Capitalist America* (New York: Basic Books, 1976), p. 9.
34. G. I. Maeroff, "Teaching Job Prospects for Graduates with Doctorates Reported to be Growing Worse," *New York Times*, 21 January 1976, p. 28; G. B. Kolata, "Projecting the Ph.D. Labor Market: NSF and BLS Disagree," *Science* 191 (1976): 36308—5; National Board on Graduate Education, *Outlook and Opportunities for Graduate Education* (Washington, D.C.: National Board on Graduate Education, 1975).
35. A. Shuster, "Many European Teachers Jobless as Economies Lag, Birthrates Dip," *New York Times*, 29 December 1976, p. 3.
36. P. M. Boffey, "Carter Aides Lament Research Decline," *Science* 197 (1977): 32; J. Walsh, "The State of Academic Science: Concern about the Vital Signs," *Science* 196 (1977): 1184–85; B.J.C., "NIH Budget on the Decline," *Science* 195 (1977): 375.
37. E. B. Fiske, "Columbia Shifting Graduate Studies from Full-time Doctoral Programs," *New York Times*, 5 January 1977, p. 1.
38. Id., "General-Education Courses Spreading to Graduate Level," *New York Times*, 22 December 1976, p. 26.
39. G. I. Maeroff, "Applications to Professional Schools Ebbing," *New York Times*, 26 November 1976, p. A1.
40. Id., "Value of Going to College Wins New Support," *New York Times*, 1 December 1976, p. B13; R. B. Freeman, *The Overeducated America* (New York: Academic Press, 1976); C. Bird, *The Case Against College* (New York: McKay, 1975); C. J. Hitch, University of California Twenty-Ninth All-University Faculty Conference (1975), "The Entering Undergraduate Student: Changes and Educational Implications," University of California at Davis.
41. L. Buder, "Schools Open Today in New York, in Tense Mood of Austere 'Crisis,'" *New York Times*, 13 September 1976, p. 1; "Closed Schools in Oregon District Pose Dilemma for the Taxpayers," *New York Times*, 8 November 1976, p. 18; W. K. Stevens, "56,000-Pupil School System Shut as Toledo Voters Bar Tax Rise," *New York Times*, 11 December 1976, p. 25; R. A. Stuart, "Cleveland Schools in Financial Limbo," *New York Times*, 8 April 1978, p. 9.
42. Perkins, *Rural Small-Scale Industry in . . . China*, p. 46.
43.. D. S. Elgin and R. A. Bushnell, "The Limits to Complexity: Are Bureaucracies Becoming Unmanageable?" *The Futurist* 11 (1977): 337–49.
44. J. T. Wooten, "Over Half in a Poll Feel Distrustful of Government," *New York Times*, 24 February 1976, p. 1.

45. Heilbroner, *The Human Prospect*.

46. D. Carmody, "Moves at International Parleys in '76 Seen as Threatening Free Press," *New York Times*, 22 December 1976, p. 12.

47. G. Fowler, "Beame's Plan Proposes 52 Districts for Delivery of Most City Services," *New York Times*, 16 September 1976, p. 52; id., "New York City's Community Boards Growing in Power," *New York Times*, 31 May 1976, p. 6.

48. T. Wicker, "Who Spent Your Money for What?" *New York Times*, 20 April 1976, p. 35.

49. Isaacs, *Idols of the Tribe*; Perkins, *Rural Small-Scale Industry in . . . China*; "Even U.S. Has a Separatist Movement as the Trend toward Fragmentation Threatens Many Countries," *New York Times*, 3 January 1977, p. 7.

50. O. Handlin, quoted in R. A. Mohl, "The Industrial City," *Environment* 18 (1976): 28–38 (quote from p. 34).

INDEX

absorption, 11–12
Accounting Office, U.S., 55–56
actualization, 279
advertising, 277
Africa, 22, 102, 280–81
age, labor and, 269, 271
Age of Consumption, 126–27
Age of Wood, 124
age structure of populations, 220–23
agribusiness, 175–76, 208
agricultural equation, 191, 192–93
agricultural industries, 114, 122, 158,
    175–84, 205–8
  in resource chain, 83, 87
Agricultural Revolution, 168–69
agriculture, 22–24, 26–27, 167–214,
    267, 271
  air pollution and, 110, 112
  carrying capacity and, 226
  chemicals in, 172, 173–74, 176, 179,
    181, 182, 199
  Chinese model in, 203–4, 226, 228
  decentralization of, 193
  demechanization of, 199–200
  despecialization of, 198–99
  development of, 22–23
  diminishing returns in, 185–87
  fossil fuels in, 168, 172, 174, 175, 178,
    181, 182, 187, 191, 193, 195, 196,
    199, 268
  Genesis Strategy for, 209–10
  Green Revolution in, 181, 205–8

  in growth of cities, 229–31, 232
  labor-intensive, 200, 204, 207–8,
    229, 254
  land policies and, 208–9
  no-till farming in, 173–74
  return to past and, 202–3
  slash-and-burn, 22–23
  stable, 23–24
  tax reform and, 208
  traditional, 181, 182
  transportation and, 169, 171,
    183–85, 193, 194
agriculture, U.S., 167–76, 182–85
  capital phase of, 169–70
  energy phase of, 170–72
  energy use in, 170–76, 182–85
  history of, 167–72
Agriculture Department, U.S., 174,
    254
air pollution, 11, 109–12, 147–48, 281
  agriculture and, 110, 112
  plant productivity and, 109–10
albedo, 11, 101, 102
Allied Chemical Company, 108
Amazon Basin, 123
Amenhotep, xvii, xxi
Andrews, P. W., 132
animal resources, 121–23, 199–200
antibiotics, 183
Arab nations, 138, 155, 205
Arndt, R. A., 128, 130, 131
asbestos, 109, 112, 116, 269

Asia, 228
atmosphere, 109–12
  pollution of, 11, 109–12, 147–48, 281
Atomic Energy Commission, 57
atrazine, 173
Australia, 193, 205
automotive industry, 60, 71
awareness, 279

bankruptcy, 237, 243, 245, 251
Bell, Daniel, xix, 133
bell-shaped model, 262
  for fossil-fuel production cycles,
    42–46
  for metal depletion, 128–32
Benstock, M., 105
biomass, 14, 69, 125
  of animals vs. plants, 18
  in food chain, 81–82
  of humans vs. animals, 196
birth rate, 218, 220, 225
Bowles, S., 272
Brazil, 115
Brooks, D. B., 132
bureaucracies, 274
Bureau of Land Management, U.S.,
  160
Bushmen, !Kung, 22

calories, 194–95, 202, 226
  defined, 18n
Canada, 56, 57, 67, 100, 124, 193, 205,
  245
cancer, 116, 122, 161, 173, 269
capitalism, 269, 270, 277
carbon dioxide:
  forest depletion and, 125
  greenhouse effect and, 102, 110–12,
    125, 161, 281
Caribbean, 207, 228
carnivores, 19–20, 81, 82, 84, 89
carrying capacity, 215–16, 224–26, 228,
  267
  consequences of, 233–34
  defined, 225
  global, 225–26
Carter administration, 72, 209, 210
  Energy Program of, 54n, 63, 70, 151
cells, 14

Census Bureau, U.S., 221, 231, 248
Central America, 207
Central Intelligence Agency (CIA), 103
cesium, 106
change:
  agriculture and, 191–214
  stability vs., 152–53
chemical energy, 5–6
chemicals, in agriculture, 172, 173–74,
    176, 179, 181, 182, 199
Cheveron Chemical Company, 174
China, 46, 112, 210, 254, 270, 273,
    280–81
  agriculture in, 185, 198, 203, 226,
    228, 267
cities, 215, 229–34, 237–61
  decline of, 237–61, 267
  dismantling of, 254
  fiscal strain in, 241–43
  formation of, 229–31
  future of, 251–52
  high cost of living in, 240–41
  in industrial vs. nonindustrial na-
    tions, 230–31, 233
  metabolism of, 232–33
  modern, 230–31
  national urban policy for, 252–54
  plight of, 237–47
  space, 265–66
civilizations, cyclical nature of, 262
civilizations, industrial, see industrial
    civilization
class, 270, 272
climate, 12–13, 98–103
  fluctuations of, 99–101
  of future, 102–3
  recent trends in, 101
climate modification, 11, 101, 102,
    111–12, 125, 161, 281
climatology, defined, 99
coal, 25–26, 33, 111, 150, 161
  bell-shaped model for, 44
  exponential model for, 38–39
  future of, 38–39, 44, 49
  gasification of, 56
  liquification of, 55–56
  synthetic fuels from, 54–56
Coca-Cola, 278
commercial sector, energy use in, 36

Commoner, Barry, 87, 95
communism, 270, 277
community:
    decline of, 264
    rebirth of, 266–68
control theory, 144–64
copper, energy upgrading and, 82–84
Cornell University, 254
Cornucopians, 132–36
cost of living, 240–42
Crawley, G. M., 7, 67
crops, 176–81
    as monocrops, 178–81, 268
    as supercrops, 176–78, 179, 208
    see also agriculture
cultural evolution, xx, 1, 20–26, 153
    natural law and, 20–21
culture, 276–80
cycles, 19, 262–63, 279, 280, 281, 282
    see also food cycle; natural resource
        cycle

DBCP, 122
DDT, 106
death rate, 218
debt service, 243
defense, 210, 273–74
democracy, 152, 153, 254, 274, 275
demographic transition, theory of,
    218–20
Depression, Great, 93–94, 134, 209,
    251
desertification, 114
dictatorships, 153
diet, changes in, 194–97
differential inflation, price structure
    and, 201–2
diminishing returns, in agriculture,
    185–87
diminishing returns, law of, 134–36,
    274
diversity, decline of, 278
double-cropping, 204
Dow Chemical Company, 122
drought, 102, 103–4, 162, 207, 224
DuPont Company, 108

earth:
    geological history of, 10–11
    wobble of, 100

Easter egg hunt (model), 40–42, 128,
    157–58
ecocultural history, lessons of, 26–31
eco-logic, birth of, 279–80
ecological pyramid, 82, 88
ecology, 20, 280
    carrying capacity in, 215–16
    Keynesian economics and, 94
    of monocropping, 179–81
    of natural resources, 77, 81–96
economic limits to industrialism,
    153–56
economics:
    Keynesian, 94
    resource cycle and, 86–96
    of resource recovery, 133
education, 88, 208, 211, 250, 271–73
efficiency, 203, 228
    of green plants, 17–18
    of herbivores vs. carnivores, 19
    productivity vs., 226
    of U.S. food system, 185, 202
Egyptians, 168
Einstein's equation ($E = MC^2$), 58, 82
electricity, 60, 64, 68, 69, 182, 233
electric utilities, energy use in, 37
electron-beam fusion, 65n
energetic efficiency, defined, 202
energy, 3–16
    in agriculture, 170–76, 182–85
    as basis of industrialism, 33–53
    in cities, 233–34, 238, 240–41, 242,
        252–54
    defined, 4
    earth's sources of, 10–11
    future of, 72–73
    future price of, 155–56
    kinetic, 4–5, 6
    for living forms, 17–18
    as natural resource, 77–78
    potential, 4–5
    quality of, 8
    in resource cycle, 82–85, 91, 94
    short-term solutions and, 50–51
    storage of, 5–6, 27–28, 31, 69
    subsidization of, 28
    upgrading of, 17–20, 22–27, 29–31,
        83–85, 194
    uses of, 35–37

energy (continued)
  see also specific types and sources of
    energy
energy conservation, 70–72
  law of, 6–7, 9, 14, 137
energy conversion, 33–35, 69, 205
  GNP and, 91, 92
  per capita, 24, 29
energy crisis, solution to, 72–73
energy degradation, law of, 7–8, 9, 14
  in food cycle, 20, 21, 84
  see also entropy
energy depletion, 1, 116, 221
  economic consequences of, 200–202
Energy Research and Development
    Administration, U.S., 55, 70, 156
England, 28, 50, 60, 87, 185, 203, 245
  Industrial Revolution in, 25, 33, 230,
    270
Enrico Fermi plant, 60
entropy, 7–8, 9–10, 77
  defined, 7
  food cycle and, 20, 81, 82
  life and, 14–15
  in natural resource chain, 84
  solar radiation and, 14–15
environmental limits, negative feed-
    back and, 146–48
Environmental Protection Agency,
    U.S., 105, 108, 151
erosion, 114
Eskimos, 225–26
ethics, 276–80
Europe, 60, 70–71, 98, 102, 104, 110,
    116, 124, 178, 207, 228, 267
exponential model:
  for fossil-fuel production cycles,
    37–42
  for metal depletion, 127–28
extractive industries, 82, 83, 84, 87, 90
extrapolation, defined, 38

family, 249, 264
family farms, 169, 175–76, 208
famine, 102, 205, 207, 224
farms, 169–76
  family vs. corporate, 175–76
  methodology of, 172–74

population of, 168, 170, 174
  size of, 174–75
  urban, 254
fast-breeder reactors, 59–63, 66
  cost of, 60
  drawbacks of, 60–62
Federal Energy Administration, U.S.,
    151
Federal Power Commission, U.S., 155
Federal Reclamation Act of 1902, 208
fertilizers, 114, 138, 179, 182, 191,
    197–99, 204, 206
  farm output and, 171–72, 177, 185
  nitrogen, 125, 197–98
fiscal strain, positive feedback and,
    241–43, 249
fiscal strength, study of, 244
fisheries, 122–23
fission, 6, 58–63
food:
  in global perspective, 205
  growing of, 182
  preparation of, 183–85
  processing of, 182–83, 194
  production of, 216–17, 224
  see also agriculture; crops
Food and Drug Administration, U.S.,
    183
food chain, 19–20, 173
  diet and, 194–96
  ecological pyramid and, 82, 88
  natural resources and, 81–84
food cycle, 19–20, 21, 26–27, 86, 121,
    191, 229–30
  entropy and, 20, 81, 82
  natural resources and, 84–87
Food for Nought (Hall), 183
food prices, future of, 200–202
food reserves, 209–10
Ford Foundation Energy Project, 154
forest resources, 123–26
  depletion of, 124–25, 228
  future of, 125–26
  use of, 123–24
Forest Service, U.S., 124, 125
Fortune magazine, 221
fossil-fuel production cycles, 37–45
  bell-shaped model of, 42–46
  exponential model of, 37–42

fossil fuels, 5–6, 14, 33, 37–51, 106,
    125, 128–29, 210
    in agriculture, 168, 172, 174, 175,
        178, 181, 182, 187–191, 193, 195,
        196, 199, 268
    alternatives to, 54–76
    climate affected by, 102
    in energy subsidization, 28
    ocean contamination and, 106, 107
    in population growth, 215
    "quality" of, 8
fragmentation, 264, 268–69, 270
France, 60, 62, 245
"free" nations, 274, 275
frequency histograms, 221–23
Freud, Sigmund, 152
Friedman, Milton, 244
Fuller, Buckminster, 267
fusion, 6, 58, 63–67
    advantages of, 63–64
    obstacles to, 64–67
    techniques for, 65–66
fusion reactors, 66–67
future, 31, 262–85
    of cities, 251–52
    climate in, 102–3
    of energy, 72–73
    of forest resources, 125
    of global population, 223–24
    of industrialism, xvii–xxi, 49, 77, 88,
        89, 262–85
    of metals, 127–32, 140
    prediction of, xviii–xx
    visions of, 262–85
    of water, 105–6
future shock, defined, 86

Galbraith, J. K., 90
Gallup poll, 238
General Electric Company, 107
Genesis Strategy, for agriculture,
    209–10
genetics, in agriculture, 176–78
Geological Survey, U.S., 49
geothermal energy, 10–11, 67–68, 70
Germany, 50, 55
Gintis, H., 272
Golden Rule, 280
goods and services, 90–91

government, central vs. local, 273–76
Great Britain, see England
greenhouse effect, carbon dioxide and,
    102, 110–12, 125, 161
green plants, 17–20
    efficiency of, 17–18
    see also biomass
Green Revolution, 181, 205–8
    redefining of, 205–7
    U.S. importation of, 207–8
Gross National Product (GNP), 91, 92,
    276
ground water, 105, 114

Hall, R. H., 183
health, 81, 250
heat energy, 6, 8–10
Heilbroner, Robert, xix, 274
herbicides, 172–74
herbivores, 18, 19, 81, 82, 84, 89
homeostasis, institutional, 152–53
honeybee population, 122
Hubbert, M. King, 44, 47–50, 128, 224,
    252
human nature, 264
hunter/gatherer societies, 21–22, 191,
    202, 277
hydrocarbon bond, 5
hydrocarbon contamination, 106, 107
hydrogen in fusion, 63–66
hydrologic cycle, 103, 106, 262

India, 181, 206, 225
Indians, American, 150, 226, 267
Industrial Age, xx, 4, 126, 159, 237,
    276, 278
    birth of, 24–26
    ending of, xviii, 211
    fragmentation in, 264
    knowledge and, xviii, xx, 203, 280
    population growth in, 31, 215
industrial civilization, 1, 121, 122, 187,
    191
    life cycle of, 263
    mineral depletion and, 126–27
industrial equation, 3, 4, 77
industrialism:
    climate modification in, 11
    defined, 3

industrialism (*continued*)
  energy basis of, 33–53
  energy conservation and, 70–72
  entropy and, 9–10
  evolution of, 17–32
  future of, xvii–xxi, 49, 77, 88, 89,
    262–85
  high price of, 153–55
  interrelatedness of limits, 159–62
  knowledge and, 157–59, 268–69
  limits to growth of, 144–64
  metals and, 127
  negative feedback in, 145–62
  political differences vs. structural
    similarities in, 270
  population growth in, 218
  stages in development of, 87–89
  uniqueness of, 29
  water in, 104
  work in relation to, 4
Industrial Revolution, xviii, xx, 25, 33,
    110, 168, 191, 216, 218, 230, 268,
    276
inflation, 91–95, 201–2, 221
intentional communities, 267–68
intercropping, 204
Interior Department, U.S., 34, 35, 57
interrelation, 167
  first-order, 159–60
  of limits to industrialism, 159–62
  nth-order, 160–62
  second-order, 160
Iran, 46
Iraq, 115
irrigation, 204, 208
isolation, defined, 100
Israel, 267
IT&T, 176

Japan, 33, 50, 138, 185, 228, 245, 248
Jefferson, Thomas, 193
job losses, in decline of city, 239–40,
    253
jobs, 87–89, 254
Joseph of Egypt, xvii, xxi, 210
Joule, James Prescott, 6–7

Kahn, Herman, 133, 157
karma, 280

Kepone, 108
kerogen, 57
kerosene, 25
Keynes, John Maynard, 94, 96
Keynesian economics, 94
kibbutzim, 267
kinetic energy, 4–5, 6
knowledge, xviii, xx, 88, 157–59, 203,
    268–69, 272, 280
Kuhn, Thomas S., xix
!Kung Bushmen, 22

labor, 200, 204, 210, 230, 254, 267, 269,
    271
  in agricultural equation, 191, 192,
    193
labor-intensive practices, 200, 204,
    207–8, 229, 254
labor market, 269
land, 112–15
  arable, 112–14, 115, 226–28, 265
  degradation of, 114–15
  future of, 115
land-ownership patterns, 150, 208–9
"land races," 177
laser-beam fusion, 65–66
Lawson's criterion, 65
Leach, G., 174
lead time limit, 148–50
Leibniz, Gottfried Wilhelm von, xx
lifeboat ethic, 210
limited processes, 144–64
  interrelatedness and, 159–62
  nature of, 144–45
Limits to Growth (Meadows et al.), 37,
    127, 128
livestock, 195–96, 197, 198–99
living patterns, 264–68
Lovins, A. B., 150

MacDonald's, 278
magnetic confinement, 65
malnutrition, 218
Malthus, Thomas, xvii, 216, 223, 224,
    228
manufacturing industries, 251, 253,
    268, 270, 271
  cities and, 230–31, 239–41, 251, 253
  in resource chain, 83, 90

manure, 198–99, 200
Marx, Karl, 269, 270
Marxists, 269–70
materialism, 277, 279–80
Mediterranean Basin, 178
Mendel, Gregor, 176
mercury contamination, 108
metals, 182
    bell-shaped depletion model for, 128–32
    depletion of, 127–32
    exponential depletion model for, 127–28
    future of, 127–32, 140
Mexico, 115, 160, 193, 207
Middle East, 51
migration, 218, 238, 241–43
military services, 254
mineral resources, 126–40
    foreign dependence and, 138–40
    substitution of, 137
Mobil Oil Company, 49
money:
    city borrowing of, 243
    resource cycle and, 90–94, 154
    value of, 90
    see also inflation
monocrops, 178–81, 268
    age of, 178–79
    ecology and, 179–81
Morocco, 138
mortality, 218
Muller, T., 241

natality (birth rate), 218, 220, 225
National Academy of Sciences, 49, 180
national defense policies, 210
National Petroleum Council, 49
national urban policy, 252–54
natural gas, 48, 111, 136, 151, 155, 156
    bell-shaped model for, 45–46
    exponential model for, 40
    future of, 40, 45–46, 49
natural law, cultural evolution and, 20–21
natural resource chain, 82–84
    defined, 82
natural resource cycle, 84–96
    economics and, 86–96

energy in, 82–85, 91, 94
    inflation and, 91–94
    money and, 90–94, 154
    positive feedback in, 85–86
natural resource depletion, 1, 124–25, 127–40, 221
    cities and, 252–54
    dynamics of, 41–43, 128
    economic consequences of, 200–202
    models of, 37–46, 127–32
natural resource pyramid, 88–89
natural resource recovery, 133, 134
natural resources, 77–143
    classification of, 78–81
    defined, 77–78
    ecology of, 77, 81–96
    food chain and, 81–84
    future of, 115–16
    jobs and, 87–89
    primary, see primary natural resources
    renewable vs. nonrenewable, 78–79
    secondary, 79, 80, 121–43
    tertiary, 79–81, 121
    see also specific resources
nature, human place in, 278–79, 280
negative feedback, 145–62
    in control theory, 145–46
    defined, 40
    environmental limits and, 146–48
New Economics, 94–96, 129
New York City:
    Community Boards of, 275–76
    urban decline in, 248–51
New Zealand, 67
"night soil," 198
nonindustrial nations:
    agriculture and, 203–4, 206–8, 209
    population in, 218–20, 221, 230–31, 233
nonrenewable resources, 78–79
North America, 98, 101, 135, 205
no-till farming method, 173–74
nuclear energy, 6, 8, 35, 58–67, 148–49, 154, 161, 266
    contamination from, 106
    energy subsidization and, 28
nuclear pirating, 61
nuclear reactors, 59–63, 66–67

Nuclear Regulatory Commission,
    61–62

occupational carcinogenesis, 269
oceans, 106–8
Odum and Odum, 95
oil, 25–26, 111, 153–54, 155–56, 184,
    205, 232
  bell-shaped model for, 44–45
  exponential model for, 39–40
  future of, 39–40, 44–45, 46–49
  human energy equivalent of, 174
  in shale and tar sands, 56–57, 160
oil-depletion allowance, 154
oil industries, 47–49, 152, 153–54
oil spills, 106
optimism, xx, 281–82
orchid economy, 93–94, 95
organic farming, 214
Organization of Petroleum Exporting
    Countries (OPEC), 47, 50, 155
outmigrants, characteristics of, 238
Owen, O. S., 78

Pakistan, 61n, 115
paradigm, xix–xx, 122
paraquat, 174
pattern, role of, xviii
pesticides, 122, 176, 179, 180–81, 185,
    199
petroleum, see oil
photosynthesis, 5–6, 13–14, 17–18, 158
  algae and, 106
photovoltaic systems, 68
phytoplankton, 106
Pilgrims, 169
Pimentel, D., 198, 199
plant productivity, air pollution and,
    109–10
Plato, 264
plutonium, 59, 60, 61–62, 106
Poland, 25
political institutions, decentralization
    of, 273–76
political limits to industrialism, 151–53
pollution, 265, 270
  air, 11, 109–12, 147–48, 281
  radioactive, 60–62, 106, 108
  water, 104–5, 106–8, 122, 281

polychlorinated biphenyls (PCBs),
    107–8
population, 215–36
  age structure of, 220–23
  demographic transition theory and,
    218–20
  future of, 223–24
  history of, 216–18
  redistribution of, 216
  urbanization of, 229–34
population densities, 226–28, 231–32,
    237
  worldwide statistics on, 227
population growth, 23, 29, 114, 215,
    216–20, 223–25
  food production and, 216–17, 224
  negative feedback and, 145–46
  positive feedback and, 31, 145, 146,
    220
positive feedback, 27, 152, 153, 169,
    230, 262, 270
  in bell-shaped model, 42–43
  in control theory, 145
  defined, 23
  in exponential growth, 40
  fiscal strain and, 241–43, 249
  population growth and, 31, 145, 146,
    220
  in resource cycle, 85–86
  role of, 29–31
potential energy, 4–5
"poverty line," 202
preindustrial cultures, 21–24
preservatives, 183
Price-Anderson Act, 154
price structure, differential inflation
    and, 201–2
primary natural resources, 79, 80,
    98–120, 121
  in agricultural equation, 191, 192–
    93
private enterprise, 154
probability, role of, xviii
production, 269–70, 271
productivity, 203, 226, 233
  measures of, 202
profit, 277
proteins, 195–97
purchasing power, 202

radioactive wastes, 106, 108
recycling of resources, 137–38
reflection, 11
renewable resources, 78–79
residential sector, energy use in, 36
resources, *see* natural resources
retirement, 221, 269
retribalization, 267
reurbanization, theory of, 247–48
Robbins, W., 196
Rocky Flats Nuclear Weapons Plant, 61
Roper, L. D., 128, 130, 131
rural communities, 267

Sahara Desert, 114
Sahel, 102, 207
salinization, 114–15, 160
Saudi Arabia, 47, 112
schooling, restructuring of, 271–73
Scotland, 245
secondary natural resources, 79, 80,
    121–43
self-development, 279
service economy, 88–89, 239–40,
    249–50, 251, 268–69, 271
shale rock, 56–57
slash-and-burn agriculture, 22–23
"Small is Beautiful," 96
Smith, Adam, xvii
Social Security system, U.S., 221, 269
social upheaval, 253–54
sociological limits to industrialism,
    150–51, 152–53
solar energy, 10–15, 68–70, 149–50,
    200
    advantages vs. disadvantages of,
        68–69
    agriculture and, 22–24
    collection of, 68
    conversion of, 69
    in hunter/gatherer societies, 22
    "quality" of, 8, 9, 14, 69
    stored energy vs., 28
South Africa, 138
South America, 280–81
Soviet Union, 46, 60, 185, 210, 245–46,
    254, 270
soybeans, 196, 197
Space Agency, U.S., 266

space cities, 265–66
specialization, 268–69, 270, 272,
    276–77, 279
Standard Oil, 176
status, in industrial societies, 277
stock market, 270
strip mining, 114
substitution of resources, 137
suburbanization, 247–48, 264, 266
sun, 10, 14–15
supercrops, 176–78, 179, 208
    defined, 176
    evolution of, 176–78
    genetic loss in, 177–78
surface water, 104–5
Sweden, 70, 71
synthetic fuels, 54–56

tar sands, 56, 57
taxes, 221, 241–43, 244, 247, 249,
    274
tax reform, agriculture and, 208
technology, 137, 253, 280
    limits of, 156–59
    of resource recovery, 134
Technology Assessment, U.S. Office of,
    134
television, 277, 278
terrorist organizations, 61
tertiary natural resources, 79–81, 121
thermal efficiency, 10
thermodynamics, laws of, 1, 6–8, 17,
    18, 19, 194
    in biology, 20
    first, 6–7, 9, 14, 137
    second, *see* energy degradation, law
        of; entropy
tidal energy, 10–11, 67–68, 70
timber industry, 123–24
time, concepts of, 276
time limits to industrialism, 148–50
"time's arrow," 7
TNT, 58
transportation, 36, 104, 151, 154, 250,
    279
    agriculture and, 169, 171, 183–85,
        193, 194
    energy conservation and, 70–71, 193

trophic levels, 19, 81–84, 194
　knowledge and, 203
　in resource chain, 84, 88–89

unemployment, 249, 254, 269, 273
United Nations, 34, 106, 114, 178, 210
United Nations World Health Organi-
　zation, 269
United Nations World Population Con-
　ference (1974), 223
United States, 148–52
　age structure of, 221–22
　agriculture in, see agriculture, U.S.
　arable land in, 112–13, 114, 115
　as breadbasket, 205, 208
　building codes in, 150–51
　calorie sources in, 195
　coal in, 38, 44, 49, 55–56, 150, 151
　energy budget of, 154, 159
　energy conservation in, 70–72
　energy dependence of, 33, 210
　energy future of, 38–40, 44–45,
　　48–51, 55–57, 70
　energy sources of, 34–35
　energy use in, 35–37
　food prices in, 201
　forest resources in, 123–24, 125–26
　legal system of, 152, 153
　metal depletion in, 129, 130–32, 134,
　　155, 159–60
　mineral resource consumption in,
　　126–27
　natural gas in, 40, 45–46, 48, 49, 136,
　　151, 155, 156
　nuclear energy in, 60–63, 67,
　　148–49, 154
　oil in, 25–26, 39–40, 44–45, 48–49,
　　56–57, 134, 136, 151, 154, 160,
　　182
　population densities in, 232
　solar energy in, 69, 70, 149–50

transportation in, 151, 154, 169, 171,
　183–84
urban decline in, 237–45, 246,
　247–51
urbanization in, 229, 231–32, 233–34
universe, 7, 10, 262, 281
uranium, 58–59
urban, defined, 231
urban decline, 237–61, 267
　economic impact of, 248–49
　human cost of, 248–51
　physical signs of, 250
　as repetitive cycle, 251
　social impact of, 249–50
urban exodus, 238, 241–43
urbanization, 114, 229–34
U.S.S.R., see Soviet Union

values, 276–80

water, fresh, 103–6, 232–33
　depletion of, 104–5, 228
　future of, 105–6
　uses of, 103–4
water table, 104–5
Water Wasteland (Zwick and Benstock),
　105
watts, defined, 112
wealth-power cycle, 152, 153
Weinberg, A. M., 63
welfare, 244
wildlife species, extinction of, 23,
　121–22
Wisconsin Ice Age, 100
wisdom, 280
work, 268–73
　defined, 4
　heat and, 8–10
　revaluation of, 268–71

Zwick, D., 105